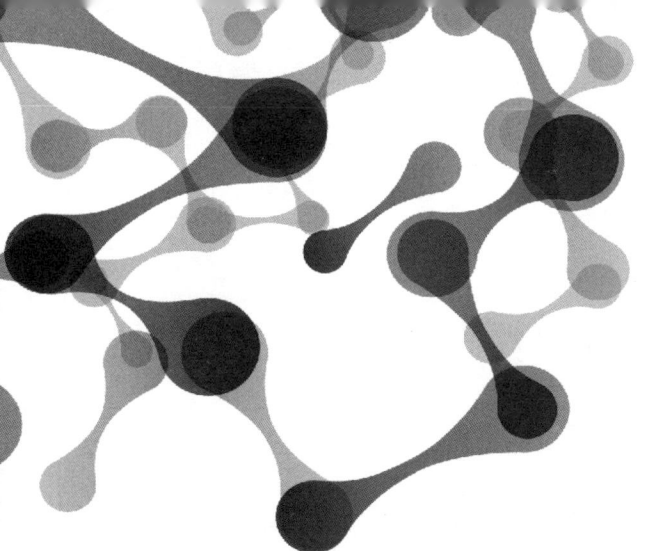

生物科学综合实验

Integrated Experiments for Biosciences

主　编　张　雁

副主编　孙彩云　何祖勇

中山大学出版社
·广州·

版权所有　翻印必究

图书在版编目(CIP)数据

生物科学综合实验/Integrated Experiments for Biosciences/张雁主编；孙彩云，何祖勇副主编. —广州：中山大学出版社，2020.8

ISBN 978 - 7 - 306 - 06898 - 9

Ⅰ. ①生… Ⅱ. ①张…②孙…③何… Ⅲ. ①生物学—实验—高等学校—教材—英文 Ⅳ. ①Q - 33

中国版本图书馆 CIP 数据核字(2020)第 121454 号

SHENGWU KEXUE ZONGHE SHIYAN

出 版 人：王天琪
策划编辑：陈　慧　曾育林
责任编辑：曾育林
封面设计：曾　斌
责任校对：马霄行
责任技编：何雅涛
出版发行：中山大学出版社
电　　话：编辑部 020-84110283，84113349，84111997，84110779
　　　　　发行部 020-84111998，84111981，84111160
地　　址：广州市新港西路 135 号
邮　　编：510275　　　　　传　真：020-84036565
网　　址：http://www.zsup.com.cn　　E-mail：zdcbs@mail.sysu.edu.cn
印 刷 者：广州市友盛彩印有限公司
规　　格：889mm×1194mm　1/16　11.875 印张　270 千字
版次印次：2020 年 8 月第 1 版　2020 年 8 月第 1 次印刷
定　　价：58.00 元

如发现本书因印装质量影响阅读，请与出版社发行部联系调换

编委会

主　　编　张　雁

副 主 编　孙彩云　何祖勇

参编人员　（按姓氏拼音首字母顺序排序）

　　　　　　曹永长　陈笑霞　黄盛丰　何祖勇

　　　　　　卢湘婉　孙彩云　张　雁　张玉婵

MPR出版物链码使用说明

本书中凡文字下方带有链码图标"＝＝＝"的地方，均可通过"泛媒关联"的"扫一扫"功能，扫描链码获得对应的多媒体内容。

您可以通过扫描下方的二维码下载"泛媒关联"APP

《生物科学综合实验》
说课（字幕版）

Preface

The textbook, "Integrated Experiments for Biosciences," is designed for senior undergraduates and graduate students. On the one hand, it plays an important role in connecting what students have previously learned with new knowledge. On the other, it is a "training" for undergraduate students before completing their graduation thesis. The textbook enables students to develop comprehensive and high-quality practical skills for acquiring biological knowledge and solving problems independently while completing integrated experiments, and thus improves the ability of students to develop good habits of deeper learning and enterprise learning.

The academic objective of the textbook is to enable students to understand experimental principles and acquire key biological skills. The "ability" objective is to enable students to acquire the ability to ask practical questions about life sciences, analyze and evaluate results, solve complex problems, and learn independently and holistically. The emotional objective is to teach students about teamwork, caring for others, and to abide by academic, ethical, and societal norms.

According to the three objectives of the textbook, each chapter is divided into two parts: learning and application. The learning section focuses on the biological dogma of DNA-RNA-Protein-Cell-Life, telling the stories of genes, proteins, cells, and life. In the application section, vivid cases are set up for students to understand the academic history and frontiers of biology, and the value of learning how biological principles and technologies are applied.

This textbook is an important learning tool to guide undergraduate students to start scientific research. The textbook is written in English, allowing undergraduates to become familiar with English academic literature and protocols from an early stage. We expect students to communicate with scholars from around the world and stand in the international academic stage, showcasing scientific contributions made by Chinese scientists to mankind.

Feb. 24, 2020, in Guangzhou

Contents

Chapter 1 DNA Isolation 1
Section 1 DNA in a Bottle 2
Section 2 Isolation and Purification of Genomic DNA from Cheek Cells 4
Section 3 DNA Isolation from Chicken Embryonic Allantoic Fluid 6
Section 4 Know More 10

Chapter 2 DNA Analysis 12
Section 1 Do You Have Jumping Alu in Your Genome? 12
Section 2 Boy or Girl? 22
Section 3 Know More 24

Chapter 3 RNA Extraction and RT-PCR 29
Section 1 RNA Extraction and Purification 29
Section 2 RT-PCR 32
Section 3 TA Cloning 34

Chapter 4 Bacterial Transformation and Plasmid Purification 38
Section 1 Bacterial Transformation and Culture 38
Section 2 Colony PCR Screening for the Reconstructed Plasmids 42
Section 3 Purification of Plasmid DNA 43
Section 4 Know More 47

Chapter 5 Protein Extraction 50
Section 1 Preparation of Protein 54

I

  Section 2 Know More ··· 55

Chapter 6 Protein Analysis ·· 60
  Section 1 SDS-PAGE and Western Blotting ·· 66
  Section 2 Know More ··· 72

Chapter 7 Animal Cell Culture ··· 75
  Section 1 Subculturing Cells ·· 87
  Section 2 Cryopreservation ··· 92
  Section 3 Recovery of Cryopreserved Cells ······································· 96
  Section 4 Know More ··· 97

Chapter 8 Cell Transfection ·· 100
  Section 1 Transient Transfection ·· 111
  Section 2 Detecting Transfection Efficiency by Flow Cytometry ············ 112

Chapter 9 Cell Contamination ·· 115
  Section 1 Types of Cell Contamination ·· 115
  Section 2 Detection of Contamination ··· 118
  Section 3 Removal of Contamination ·· 120
  Section 4 Know More ··· 122

Chapter 10 Genome Editing by CRISPR/Cas9 ································ 124
  Section 1 CRISPR/Cas9 System ··· 125
  Section 2 Cloning a gRNA into a Vector ·· 133
  Section 3 Purification of Cas9/sgRNA Coexpression Vectors ················ 136
  Section 4 Cotransfection of Cas9/sgRNA Coexpression Vectors into the Cells ········ 139
  Section 5 PCR Analysis of Targeted Deletion in the Genome ················ 142
  Section 6 Know More ··· 144

Chapter 11 Biosafety in Laboratory ·· 148
  Section 1 General Laboratory Safety ··· 148

　　Section 2　Biosafety ··· 152

Chapter 12　The Laboratory Report ··· 160

Appendix ·· 164
　　Section 1　Experimental Instruments ··· 164
　　Section 2　Preparation of Commonly—Used Solutions ································· 177

Chapter 1 DNA Isolation

▸ Background Reading

The human understanding of nucleic acids began in 1869, when Friedrich Miescher (Fig. 1-1), a Swiss doctor, isolated an extract he presumably called a "nuclein" from white blood cells from pus-filled bandages at the hospital where he worked. This nuclein was a crude extract composed of multiple proteins. Once the proteins were removed, the "pure" substance became known as deoxyribonucleic acid (DNA). It was the first time that a crude extract of DNA had been successfully extracted from cells. Interestingly, the method used at the time was quite simple; the nucleic acid was precipitated from the acidic solution simply by changing its pH. This is also the earliest description of the properties of characterizing nucleic acids, which are soluble in alkaline solutions but not in acidic ones. After this discovery, scientists found better ways to purify DNA.

Fig. 1-1 Dr. Friedrich Miescher

To study DNA, it must first be extracted from cells. This is accomplished by rupturing or lysing cells using a detergent such as sodium dodecyl sulfate (SDS). DNA-binding proteins are digested using a protease that has been added to the lysis buffer. The protease also digests DNase, which is an enzyme that would break down the DNA being harvested. Salt is added to the cell extract to enable the negatively charged DNA polymers to clump together and precipitate when ethanol is added.

Once extracted, DNA can be analyzed and studied in many experiments. Its applications include testing for ancestry, studying evolutionary relationships, paternity or maternity testing, identifying disaster victims, diagnosing and researching human genetic disorders.

Section 1 DNA in a Bottle

Objectives

- Extraction and precipitation of DNA
- Making of a DNA accessory

Materials

Reagents: Potable water; 4 × Lysis buffer; Proteinase K; ddH$_2$O; 99.9% EtOH

Supplies: Cup; Pipette tips; Microcentrifuge tubes; Ice buckets; Marker pen; Small container; Floater

Equipments: Pipette; Water bath

Activity Protocol

(1) Obtain a cup containing 3 mL water and a 15 mL tube. Label the cup and tube with your initials.

(2) Gently chew the insides of your cheeks for 30 s.

(3) Take the water from the cup into your mouth, and swish the water around vigorously for 30 s.

(4) Carefully expel the liquid back into the cup, and transfer 3 mL of the liquid into the 15 mL tube.

(5) Obtain the tube of cell lysis buffer (4 ×) from your workstation, and add 1 mL of lysis buffer to your tube.

(6) Place the cap on the tube, and gently invert your tube 5 times (don't shake it!). Observe your tube, do you notice any changes?

(7) Add 20 μL of protease K solution to the 15 mL tube containing your cell extract. Cap the tube and gently invert it again 5 times to mix.

(8) Place the tube containing your cell extract in 70 ℃ water bath for 10 min, to allow the protease K to work.

(9) Hold the tube containing your cell extract at a 45° angle and slowly add 1 volume (4 mL) of cold ethanol to the inside wall of the tube.

(10) Place the tube upright and leave it undisturbed at room temperature for 5 min. Record any observations.

(11) After 5 min, look again at the contents of the tube, especially in the area where the ethanol and cell extract layers meet. Record your observations. Compare your sample with those of your classmates.

(12) While the tube is tightly capped,

mix the contents by slowly inverting the tube 5 times. Look for any stringy or white material. This is your DNA! (Fig. 1-2, Fig. 1-3)

(13) Using a transfer pipet, carefully transfer the precipitated DNA along with the pure ethanol into the vial. Then seal the vial to complete making the accessory (Fig. 1-4).

Results

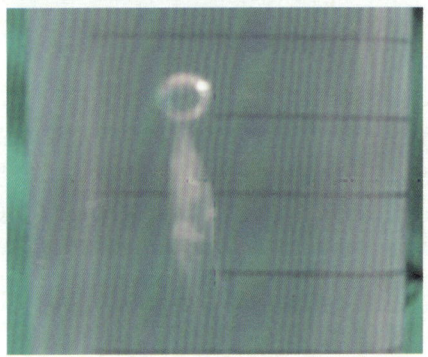

Fig. 1-3　Precipitation of genomic DNA isolated from cheek cells

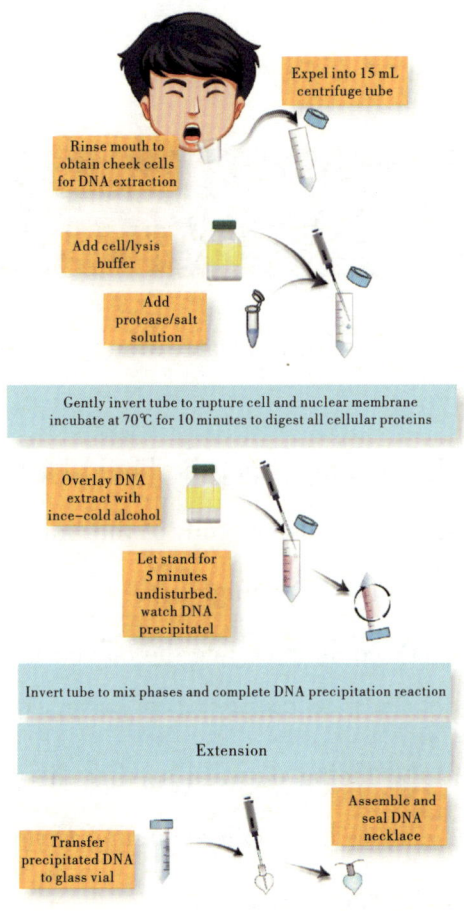

Fig. 1-2　The workflow of the experiment

Fig. 1-4　Genomic DNA in a bottle

Note: From the start of the human genome project in the 1990s to the end of the 20th century, biotechnology has been rapidly developed. Along with many important achievements in biomedical research, the acquisition, collection, storage, and application of personal biological information (e. g., fingerprints, irises, faces, and DNA) is becoming increasingly convenient and popular. However, if there are loopholes or improper procedures in the collection, the confidential use and protection of biological information will bring great challenges in controlling personal information, life and property, social governance, and even national security.

Laws should be implemented to confirm and guarantee the safety of the biological information of individuals. For the violation of the biological information of a citizen, the law should define and give punishment; the law should recognize and protect the rights and interests arising from the study and application of the genes of the individual. Discrimination and health damage caused by the leakage of biological information should be prohibited and severely punished by law, and a compensation should be paid.

Section 2
Isolation and Purification of Genomic DNA from Cheek Cells

❖ Objectives

- Collection of cheek cells
- Extraction of purified genomic DNA

❖ Materials

Reagents: MicroElute Genomic DNA Kit (Omega); Potable water; ddH$_2$O; 99.9% EtOH

Supplies: Cup; Pipette tips; Microcentrifuge tubes; Ice buckets; Marker pen; Floater

Equipment: Pipette; Microcentrifuge; Water bath; Spectrophotometer

❖ Activity Protocol

(1) Obtain a cup containing 3 mL water and a 15 mL tube. Label the cup and tube with your initials.

(2) Gently chew the insides of your cheeks for 30 s.

(3) Take the water from the cup into your mouth, and swish the water around vigorously for 30 s.

(4) Carefully expel the liquid back into the cup, and transfer 1 mL of the liquid into a 1.5 mL microcentrifuge tube.

(5) Place the tube into the microcentrifuge in a balanced configuration. Place the hinges of the tube facing outward so that the pellet will be easy to locate after spinning. Centrifuge at maximus speed for 2 min.

(6) Locate the cell pellet. The pellet should be white and be the size of a match head. If the pellet is smaller than a match head, remove the supernatant, add another 1 mL of mouth rinse to the same tube, and repeat the centrifugation.

(7) After pelleting the cells, pour off the solution. Be careful not to lose the pellet. A

small amount of saline (100 μL) should remain in the bottom of the tube.

(8) Resuspend the pellet by vortexing or flicking the tube so that no clumps of cells remain.

(9) Add 20 μL of Protease solution and mix well by vortexing.

(10) Add 120 μL Buffer BL and 4 μL of liner acrylamide. Vortex to mix and incubate at 70 ℃ for 10 min in a water bath. After 5 min of incubation, vortex or shake the tube vigorously, and then place them back in the 70 ℃ water bath for the remaining 5 min.

(11) Add 120 μL absolute ethanol to the sample, and mix again by pulse-vortexing for 15 s. After mixing, briefly centrifuge the 1.5 mL microcentrifuge tube to remove drops from the inside of the lid.

(12) Assemble a HiBind McroElute column in a 2 mL collection tube (provided), carefully apply the mixture from step 4 to the MicroElute column (in a 2 mL collection tube), including any precipitate that may have formed. Close the cap, and centrifuge at 8,000 × g for 1 min to bind DNA. Discard the collection tube and flow-through liquid.

(13) Place the column into a new collection tube (supplied). Add 500 μL of buffer HB in the column. Close the cap and centrifuge at 8,000 × g for 1 min. Discard the flow-through and re-use collection tube.

(14) Place the column into the same 2 mL tube (supplied) and wash by adding 650 μL of DNA wash buffer. Close the cap and centrifuge at 8,000 × g for 1 min. Again, dispose of collection tube and flow-through liquid.

(15) Using a new collection tube, wash the column with a second 650 μL of DNA wash buffer and centrifuge as above. Discard flow-through and re-use the collection tube.

(16) Using the same 2 mL collection tube, centrifuge empty column at maximum speed for 3 min to dry the HiBind membrane. This step is crucial for ensuring optimal elution in the next step.

(17) Place the column in a clean 1.5 mL microcentrifuge tube. Open the column and add about 20 μL preheated (70 ℃) elution buffer (water) onto the center of the membrane.

(18) Incubate at room temperature for 3 min, and then centrifuge at 12,000 × g for 1 min to elute DNA from the column (Fig. 1-5).

(19) Measure the concentration and purification using spectrophotometer (Fig. 1-6). DNA concentration = $A_{260} \times 50 \times$ (Dilution Factor) μg/mL.

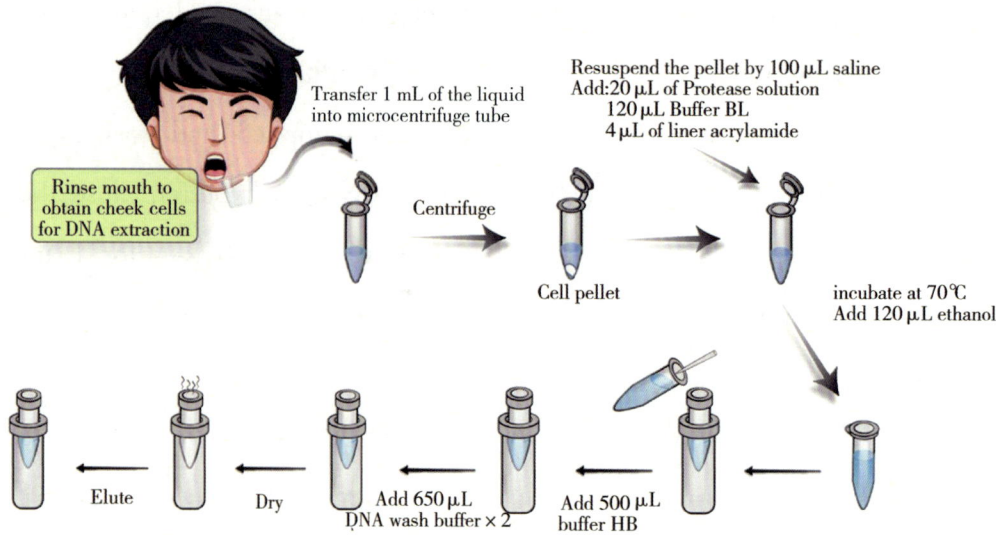

Fig. 1-5　The workflow of purified DNA isolation

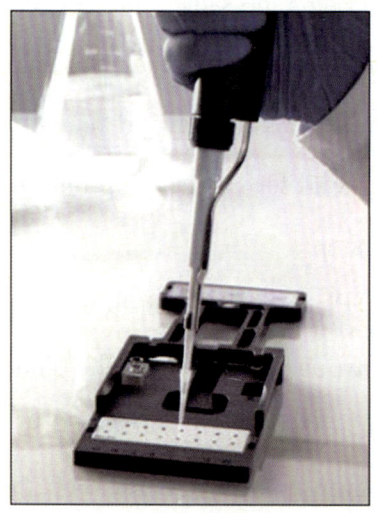

Fig. 1-6　Microplate matching with multivolume spectrophotometer

Note: *The purity of nucleic acids can be estimated by measuring the ratio of OD values at 260 nm and 280 nm (OD 260/ OD280). The ratio of pure DNA is 1.8. If the ratio is higher than 1.8, it indicates that the RNA in the DNA sample has not been removed completely. If the sample contains proteins, the ratio will be reduced.*

Section 3
DNA Isolation from Chicken Embryonic Allantoic Fluid

There are several steps involved in the formation of a regular chicken embryo. The yolk is produced by the ovary of the hen in a process called ovulation. The yolk is released into the oviduct, where it can be fertilized internally by a sperm. The yolk then moves along the oviduct (either fertilized or unfertilized) and is covered with a vitelline membrane, structural fibers, and layers of albumin (the egg white). As the egg goes down through the oviduct, it is continually rotating within the spiraling tube. This movement twists the structural chalazae, forming rope-like strands that anchor the yolk in the thick egg white. There are two chalazae anchoring each yolk on opposite ends of the egg. The eggshell is deposited around the egg in the lower part of the oviduct of the hen, just before it is laid (Fig. 1-7). The shell is made of calcite, a crystalline form of calcium carbonate.

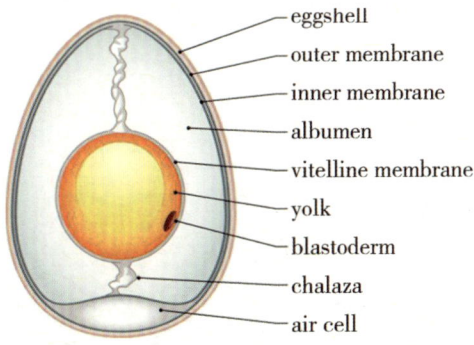

Fig. 1-7 The structure of a chicken embryo

The embryo developed inside the egg for 21 days, which is known as the incubation period, until a chick pecked its way out from its eggshell. On the first day, the blastoderm began to develop and the organ primordium emerged. On the third day, the head of the embryo was distinguishable from the tail. The visceral yolk sac and the allantois developed fast. On the seventh day, the amniotic fluid distinctively increased, and the characteristics of the chicken embryo could be seen; it has been reported that gender can be distinguished at this stage.

On the ninth day, the allantois stretched to the small end; meanwhile, the enterocoelia had been closed, and the cartilage began to ossify. On the tenth day, the allantois was developed, and feathers began to appear in the embryo.

In the following days, many organs differentiated further, and the shadow in embryos grew faster with each passing day. From the 16th to the 19th day, the chicken baby grew rapidly with feathers covered completely, and the head of the embryo had already stepped into the gas hole. On the 20th day, the lungs of the embryo began to function, and the little chicken could not wait to come out to see the world outside (Fig. 1-8).

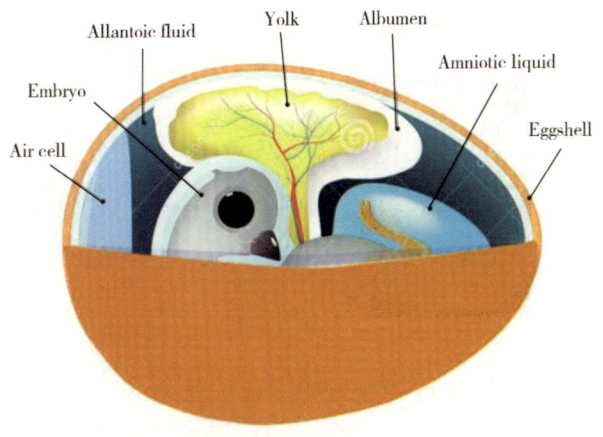

Fig. 1-8 Development of a chicken embryo

Part 1. Collection of allantoic fluid from chicken embryos

The methods used to punch the embryo and collect the allantoic fluid are important in the development of the embryo, which needs clean, careful, and harmless operations.

Objectives

- Collecting allantoic fluid
- Keeping the embryo healthy

Materials

Reagents: Iodine; 75% EtOH

Supplies: Microcentrifuge tubes; Syringe; Pencil; Marker pen; Flashlight; Candle

Equipments: 37 ℃ Incubator

Activity protocol

(1) Obtain a sterilized 15 mL tube. Label the tube with your initials.

(2) Light the embryo in dark, and at the same time, mark the position of the head air chamber by symbol "×" for punching a hole.

(3) The iodine and 75% ethanol are indispensable to sterilize. Use cotton balls which mix with 75% ethanol and iodine to clear the embryos skin, respectively.

(4) Use puncher to poke the embryo quickly on the top of gas chamber, and then poke above allantoic member with the same method.

(5) Insert the syringe into allantoic fluid slightly and extract 200-300 μL fluid.

(6) Seal the hole with melting candle.

(7) Put the embryos into 37 ℃ incubator (Fig. 1-9).

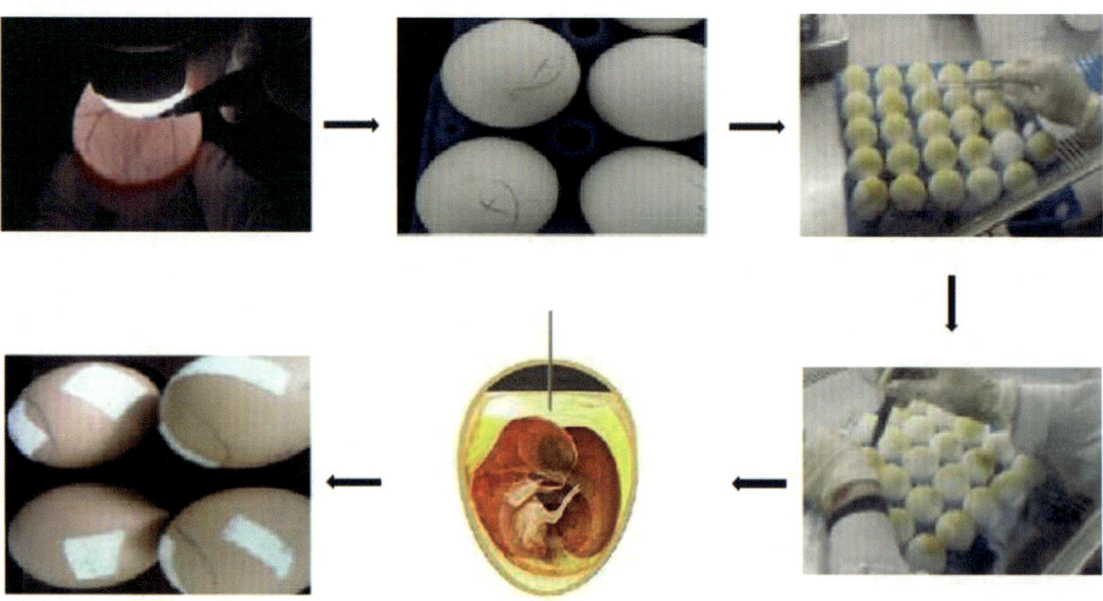

Fig. 1-9 The process of collecting the allantoic fluid from chicken embryos

Part 2. Isolation of DNA from the allantoic fluid

✥ Objectives

❖ Isolating of DNA from the allantoic fluid

❖ Measuring the concentration of purified DNA

✥ Materials

Reagents: MicroElute Genomic DNA Kit (Omega)

Supplies: Pipette tips; Microcentrifuge tubes; Ice buckets; Marker pen; Floater

Equipment: Pipette; Microcentrifuge; Water bath; Spectrophotometer

✥ Activity Protocol

(1) Label a 1.5 mL microcentrifuge tube and pipet 100 μL sample into the tube.

(2) Add 20 μL of Protease solution and mix well by vortexing.

(3) Add 120 μL Buffer BL and 4 μL of Liner acrylamide. Vortex to mix and incubate at 70 ℃ for 10 min in a water bath. After 5 min of incubation, vortex or shake the tube vigorously, and then place them back in the 70 ℃ water bath for the remaining 5 min.

(4) Add 120 μL absolute ethanol to the sample, and mix again by pulse-vortexing for 15 s. After mixing, briefly centrifuge the 1.5 mL microcentrifuge tube to remove drops from the inside of the lid.

(5) Assemble a HiBind McroElute column in a 2 mL collection tube (provided), carefully apply the mixture from step 4 to the MicroElute column (in a 2 mL collection tube), including any precipitate that may have formed. Close the cap, and centrifuge at 8,000 × g for 1 min to bind DNA. Discard the collection tube and flow-through liquid.

(6) Place the column into a new collection tube (supplied). Add 500 μL of buffer HB in the column. Close the cap and centrifuge at 8,000 × g for 1 min. Discard the flow-through and re-use collection tube.

(7) Place the column into the same 2 mL tube (supplied) and wash by adding 650 μL of DNA wash buffer. Close the cap and centrifuge at 8,000 × g for 1 min. Again, dispose of collection tube and flow-through liquid.

(8) Using a new collection tube, wash the column with a second 650 μL of DNA wash buffer and centrifuge as above. Discard flow-through liquid and re-use the collection tube.

(9) Using the same 2 mL collection tube, centrifuge empty column at maximum speed for 3 min to dry the HiBind membrane. This step is crucial for ensuring optimal elution in the next step.

(10) Place the column in a clean 1.5 mL

microcentrifuge tube. Open the column and add about 20 μL preheated (70 ℃) elution buffer (water) onto the center of the membrane. Incubate at room temperature for 3 min, and then centrifuge at 12,000 × g for 1 min to elute DNA from the column.

(11) Measure the concentration and purification using spectrophotometer (Fig. 1-10).

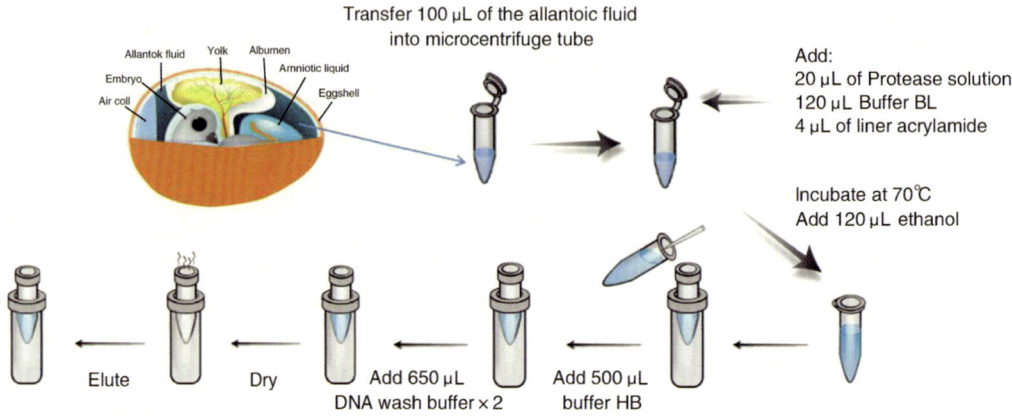

Fig. 1-10 The workflow of DNA isolation from the allantoic fluid

Section 4   Know More

DNA "hard disk"

The data stored in a computer is represented by the numbers "0" and "1", and each character and punctuation consists of a unique string of "0" and "1". Likewise, DNA stores all the genetic information of biological organisms by encoding four bases in different sequences. The potential to store information in our own genetic code has long been coveted, but how to encode the information into the DNA and how to retrieve the information has been a puzzle.

Recently, synthetic DNA bases have been found to store information instead of "0" and "1". We can assume A for 00, C for 01, G for 10, and T for 11, and a character that is supposed to be represented by 8 bits is encoded in DNA with only four bases. For example, the code for the letter "e" is 01100101. As a result, in the way of DNA, the letter "e" is encoded by a "CGCC" sequence (Fig. 1-11).

Koch and the colleagues[1] developed a storage architecture called "DNA-of-things" (DoT) for immutable memory. The researchers encapsulated the DNA in nanometer silica beads, which are fused into various materials used for printing or casting, and then 3-D printed a rabbit that contained a 45 kb digital DNA. Furthermore, a 2 min-video was stored in DNA in plexiglass spectacle lenses; the information of the video could be retrieved by the DNA embedded in a small piece of the plexiglass. The DoT approach provides a powerful storage technology, which is able to store electronic health records and precious data.

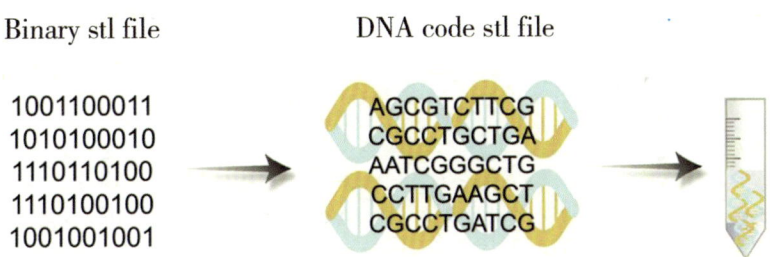

Fig. 1-11　The digital file is encoded into a DNA oligo library

≫ References

KOCH J, GANTENBEIN S, et al. A DNA-of-things storage architecture to create materials with embedded memory [J]. Nat biotechnol, 2020, 38: 39-43.

Postlab Focus Questions

1. Please write down several ways to long-term preserve DNA.

2. DNA of each person is unique. Do you know the principle of individual identification through DNA technique? How can individual identification be used?

(Written by Zhang Yan, Cao Yongchang)

Chapter 2　DNA Analysis

Background Reading

Since the discovery of the DNA double helix, the field of life science has entered an era of innovation. Genetic molecular mechanisms, including DNA replication, the genetic code, the central principle of genetic information transmission, gene expression regulation and so on, have been elucidated. As the genetic material of organisms, DNA has been studied deeply and DNA technology has also advanced rapidly as a promising tool in an array of fields. For example, DNA technologies have been applied in many areas, including medical testing, forensic identification, genetic engineering, environmental testing, agriculture, animal husbandry, food, pharmaceuticals, gene therapy, and animal cloning.

Section 1　Do You Have Jumping Alu in Your Genome?

The concept of Thomas Hunt Morgan visualizing genes as beads along the length of a chromosome barely changed throughout the first half of the 20th century. Genes were seen as inviolate objects with fixed positions on chromosomes. However, in the 1950s, Barbara McClintock showed that certain DNA fragments, termed transposons, can be activated to transpose ("jump") from one position to another. She hypothesized that transposition provides a means to rapidly reorganize genes in response to environmental stress. Her work was remarkable, not only for the fact that it flew in the face of prevailing dogma, but also because it was based entirely on the observation of chromosomes and genetic crosses. Her hypotheses had to await the discovery of modern tools of DNA analysis for confirmation. This work paved the way for the modern concept of chromosomes as dynamic, changing structures.

The Alu family of mobile genetic elements is unique to primates and are the most important short interspersed elements (SINEs). Having over 65 million years of evolution, the Alu sequence has accumulated about 1 million copies in the genome, accounting for more than 11% of the genome content. Alu is an example of the so-called "jumping gene", a transposable DNA sequence that "reproduces" by copying and inserting itself into new chromosomal loci[1]. Alu is classified as a retroposon in which its sequence structure is composed of two monomer units. The 31 bp monomer unit on the right side is longer than the left side monomer (the only active monomer) and each contains RNA polymerase III promoter sequences. The two wings of the Alu sequence have a forward repeating sequence of 7-20 bp in length, there is a poly(A) tail of dozens of bp long at the 3′-end of the Alu sequence, there is a region rich in poly(A) tails between the two monomers, and there are more CpG islands in the Alu sequence. The average spacing of Alu sequence copies is about 4 kb (Fig. 2-1).

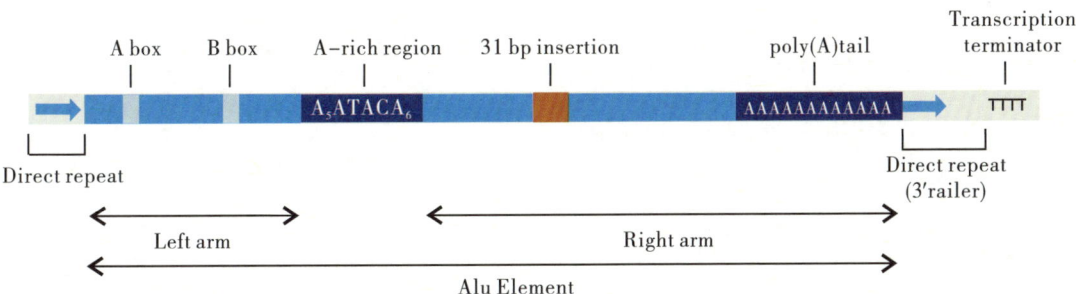

Fig. 2-1 The structure of Alu element

Alu elements can bind to many transcription factors to regulate gene expression. By analyzing Alu insertion polymorphism in the DNA of different human population or geographic locations of the same race, it can be found that Alu insertion polymorphism shows significant differences, suggesting that Alu polymorphism can be used as a stable genetic marker in human genetic research.

In this activity, the presence of the PV92-Alu repeat located in chromosome 16 is used to estimate the frequency of an insert in a class population. The presence or absence of the PV92 allele has no known correlation to disease or familial relationships. The PV92-Alu insert is 282 bp long. If the Alu sequence is absent following the amplification of the

PV92 region by PCR, the PCR product will be xxx bp. If the Alu sequence is present, the PCR product will be xxx + 282 bp (Fig. 2-2).

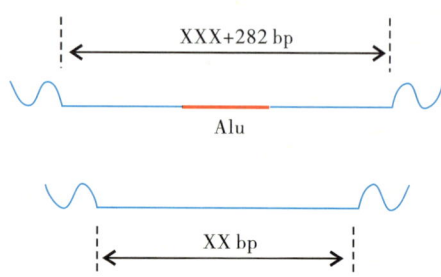

Fig. 2-2   Predicted PCR result of PV92-Alu

📖 **Information Box 2-1.** PCR is one of the most important ingenious molecular biology research tools of the 20th century that allows exponential amplification of short DNA sequences within a longer double-stranded DNA molecule. It was invented by Kary Mullis (Fig. 2-3) in association with Fred A. Faloona, Henry A. Erlich, and Randall K. Saiki in the year 1983. Mullis summarized the procedure as follows: "Beginning with a single molecule of the genetic material DNA, the PCR can generate 100 billion similar molecules in an afternoon. The reaction is easy to execute. It requires no more than a test tube, a few simple reagents, and a source of heat." In 1993, K. Mullis won the chemistry Nobel Prize for developing PCR.

Fig. 2-3   Dr. Kary Mullis

PCR is a technique based on the principle of DNA polymerization reactions. It relies on thermal cycling consisting of repeated reaction cycles of heating and cooling for melting and enzymatic replication of the DNA using a thermostable DNA polymerase, primer sequence (complementary to target region), and dNTPs. Thus, it can amplify a specific sequence of DNA by as many as one billion times. Most PCR methods can amplify DNA fragments of up to ~10 kb, although some techniques allow for amplification of fragments up to 40 kb in size.

⚙ Components

The basic components and reagents required to set up a PCR reaction are:

1. Microfuge tube and micropipette

These are small cylindrical plastic conical containers with conical bottoms with a snap cap. They are composed of polypropylene, and thus they can withstand a wide range of temperatures. Micropipettes are used to measure volumes of 1 mL or less. Some micropipettes are designed to measure fixed volumes, while others are adjustable and accommodate a variety of volume ranges. The most common volume rages are 0.5 – 10 μL, 2 – 20 μL, 20 – 200 μL, and 100 – 1000 μL.

2. Thermal cycler

It is an apparatus used to amplify segments of DNA. It has a thermal block with holes where tubes holding the PCR reaction mixtures can be inserted. The cycler works on the principle of the Peltier effect, which raises and lowers the temperature of the block in a pre-programmed manner by reversing the electric current.

3. DNA template

The reaction solution should contain at least 20 ng (20-100 ng is appropriate).

4. Primers

These are oligonucleotides that define the sequence to be amplified. Two primers that are complementary to the 3′ (three prime) ends of each of the sense and anti-sense strands of the DNA target. The GC content (the number of G's and C's in the primer as a percentage of the total bases) of the primer should be 40%-60%.

5. PCR buffer

The PCR buffer provides the optimal ionic concentration of monovalent and divalent cations and buffers to maintain the pH for optimal enzyme activity.

6. MgCl$_2$

The role of MgCl$_2$ in PCR reaction is to enhance the DNA amplification by boosting the activity of Taq DNA polymerase. It is beneficial to optimize magnesium ion concentration. The magnesium ion affects the primer annealing, strand dissociation temperatures of the template and PCR product, product specificity, the formation of primer-dimer artifacts, and enzymatic activity and fidelity. Taq DNA polymerase requires free magnesium that binds to the template DNA, primers, and dNTPs.

7. Distilled water

Autoclaved distilled water or DNase and RNase-free water can be used. The volume depends on the reaction.

8. dNTPs

These are the DNA building blocks. The dTTP, dCTP, dATP, and dGTP solutions are neutralized to a pH of 7.0. Primary stock solution are diluted, aliquoted, and stored at −20 ℃.

9. DNA polymerase

It is an enzyme used to catalyze the PCR reaction. Taq DNA polymerase isolated from Thermus aquaticus growing in hot springs functions best at 72 ℃, and the denaturation temperature of 90 ℃ does not des-troy its enzymatic activity. Other thermostable enzymes, such as Pfu DNA polymerase isolated from Pyrococcus furiosus and Vent polymerase isolated from Thermococcus litoralis, were discovered and found to be more efficient. A recommended concentration of Taq polymerase is between 1 and 2.5 units per 100 μL of reaction. However enzymatic activity will vary with respect to indivi-dual target templates or primers.

PCR is built on 20-40 repeated cycles where the temperature changes in each cycle (Fig. 2-4). The cycling starts with a single temperature step at a high temperature and is followed by one cycle held at the end for final product extension or for brief storage. The various steps of PCR are:

Step 1: Initialization

It is the first step of the cycle consists of raising the temperature of the reaction to 94-96 ℃ or 98 ℃ if extremely thermostable polymerases are used, which is held for 1 – 9 min. This process activates the DNA polymerase used in the reaction.

Step 2: Denaturation

It consists of heating the reaction to 94-98 ℃ for 20 s to 1 min. This helps in breaking of the hydrogen bonds between complementary bases, yielding single-stranded DNA molecules.

Step 3: Annealing

The mixture is cooled to a temperature of 50-65 ℃ for 20 s to 1 min, which helps in the annealing of primers to the single-stranded DNA template. Stable DNA-DNA hydrogen bonds are only formed when the primer sequence closely matches the template sequence that permits annealing of the primer to the complementary sequences in the DNA. As a rule, these sequences are located at the 3′-end of the two strands of the DNA segment to be amplified. The duration of the annealing step is usually 1 min during the first cycle as well as the subsequent cycles of PCR. Since the primer concentration is kept quite high relative to that of the template DNA, primer-template hybrid formation is favored over the re-annealing of the template strands.

Step 4: Extension

It is a DNA polymerase-dependent process. The temperature used in this step depends on the DNA polymerase used; Taq polymerase has its optimum activity temperature at 75 – 80 ℃. The temperature is adjusted so that the DNA polymerase synthesizes the complementary strands by using the 3′-OH end of the primer. The primers are extended towards each other so that the DNA segment now located between them is copied; this is ensured by employing primers complementary to the 3′-ends of the DNA segment to be amplified. The duration of primer extension is usually 2 min at 72 ℃. Taq polymerase usually amplifies DNA fragments of up to 2 kb; special reaction conditions are necessary for the amplification of longer segments. As a rule of thumb, DNA polymerase will polymerize a thousand bases per minute at its optimum temperature, leading to exponential (geometric) amplification of the specific DNA fragment.

Step 5: Final Extension

This step is performed at a temperature of 72 ℃ for 5 – 15 min after the last PCR cycle to ensure that any remaining single-stranded DNA is fully extended.

Step 6: Final Hold

In this step, the mixture is allowed to cool to a temperature of 15 ℃ for short-term storage of the reaction.

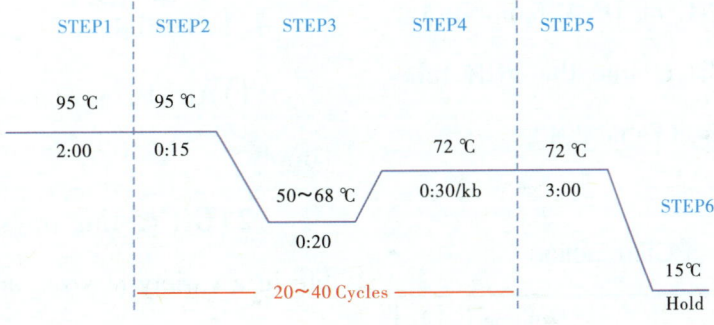

Fig. 2-4 PCR program

✿ Objectives

❖ Amplification of Alu fragment by PCR using genomic DNA isolated from cheek cells

❖ Sample loading and running of PCR products

❖ Genotype determination of DNA samples

✿ Materials

Reagents: PCR Kit; Specific primers;

DNase, RNase-free H$_2$O

Supplies: Pipette tips; PCR tubes (200 μL); Ice buckets; Marker pen

Equipment: Pipette; Microcentrifuge; Thermal cycler

Activity Protocol

Part 1. Setting up PCR reactions

(1) Centrifuge the tubes containing DNA templates at maximum speed for several seconds.

(2) Label a PCR tube with your initials and place it in the PCR tube rack on ice.

(3) Prepare master mix, containing PCR buffer, dNTP mixture, Ex-Taq, sense primer and antisense primer, on ice.

(4) Pipet 4 μL of the master mix into a PCR tube.

(5) Pipet 16 μL of DNA template (at least 20 ng) and ddH$_2$O into the PCR tube containing master mix (Table 2-1).

Table 2-1 PCR reaction

Component	Volume (μL)
10 × PCR Buffer II	2
dNTP Mixture (10 mM)	0.8
Sense Primer (10 μM)	0.5
Antisense Primer (10 μM)	0.5
Takara Ex Taq HS (5 U/μL)	0.2
DNA Template *	X
DNase and RNase Free dH$_2$O *	Y

* X + Y = 16 μL

(6) Mix by pipetting up and down 2-3 times. Cap the PCR tubes tightly.

(7) Spin down the PCR tubes for several seconds to collect the reactions.

(8) Place the PCR tubes in the thermal cycler.

Program the thermal cycler with the following program:

Initial denature: 95 ℃ for 2-5 min

35 cycles of: 95 ℃ for 30 s; 60 ℃ for 30 s; 72 ℃ for 30 s.

Final extension: 72 ℃ for 10 min

Hold: 15 ℃

Part 2. Preparing and running a standard agarose gel

1. Essentials

(1) An electrophoresis chamber and power supply.

(2) Gel casting trays, which are available in a variety of sizes and composed of ultra-violet (UV)-transparent plastic. The open ends of the trays are closed with tape while the gel is being cast, then removed prior to electrophoresis.

(3) Sample combs, around which molten agarose is poured to form sample wells in the gel.

(4) Electrophoresis buffer, usually composed of Tris-acetate-EDTA (TAE) or Tris-borate-EDTA (TBE).

(5) Loading buffer, which contains something dense (e.g., glycerol) to allow the sample to "fall" into the sample wells, and one or two tracking dyes, which migrate in the gel and allow visual monitoring or how far the electrophoresis has proceeded.

(6) SYBR Green I, an asymmetrical cyanine dye used as a nucleic acidstain in molecular biology. SYBR Green I binds to DNA. The resulting DNA-dye-complex absorbs blue light and emits green light. Some commercial products are a mixture of SYBR Green I and loading dye.

(7) Transilluminator, which is an ultraviolet light box used to visualize ethidium bromide-stained DNA in gels.

Note: Always wear protective eyewear when observing DNA on a transilluminator to prevent eye damage from UV light exposure.

Note: General Safety Requirements

(1) Always wear a lab coat and gloves.

(2) Do not talk while Eppendorf microcentrifuge tubes are open (avoid contamination).

(3) Hold pipettor with the tip facing down.

(4) Ethidium bromide is thought to act as a mutagen because it intercalates double stranded DNA (i.e., inserts itself between the strands), deforming the DNA. This could affect DNA biological processes, including DNA replication and transcription. Ethidium bromide is not regulated as hazardous waste at low concentrations, but is treated as hazardous waste by many organizations. This material should be handled according its material safety data sheet.

2. Preparing agarose gel

(1) Agarose powder is mixed with electrophoresis buffer until the desired concentration is reached.

(2) Heat in a microwave oven until the agarose is completely melted.

(3) After cooling the solution to about 60 ℃, it is poured into a casting tray containing a sample comb and allowed to solidify at room temperature (Fig. 2-5) or in a refrigerator to speed up the solidification.

(4) After the gel is solidified, the comb is removed carefully to not rip the bottom of the wells.

(5) The gel, still in its plastic tray, is inserted horizontally into the electrophoresis chamber and covered with electrophoresis buffer.

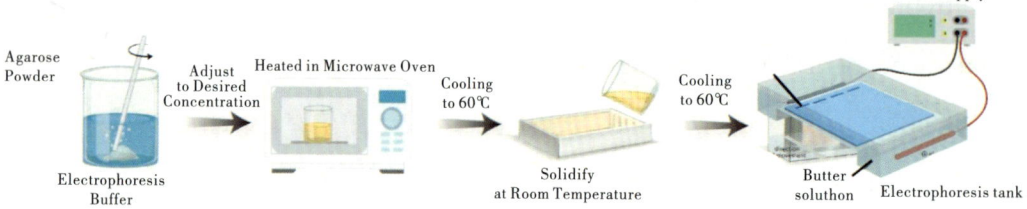

Fig. 2-5 The procedure of preparing agarose gel

3. Loading samples

(1) Pour an agarose gel and cover it with 1 × TAE buffer.

(2) Pipette 1 μL drops of the SYBR Green I (mixed with loading dye) onto a piece of plastic wrap (Fig. 2-6A, one for each DNA sample).

(3) Pipette 5 μL of each DNA sample into one of the drops of loading dye. Mix by pipetting up and down and carefully load into a well of the agarose gel (Fig. 2-6B).

(4) Pipette 6 μL of DNA marker into an empty well of the gel.

(5) Place the lid on the gel box and turn on the power supply to 100 V.

(6) After the tracking dye has migrated half to two-thirds of the way through the gel, turn off the power and remove the gel from the electrophoresis tank. Observe the gel through the UV.

Information Box 2-2: Samples containing DNA mixed with loading buffer are then pipetted into the sample wells, the lid and power leads are placed on the apparatus, and a current is applied. The current flow can be confirmed by observing bubbles coming off the electrodes. DNA will migrate towards the positive electrode, which is usually colored red.

The distance DNA has migrated in the gel can be judged by visually monitoring the migration of the tracking dyes. The Bromophenol blue and xylene cyanol dyes migrate through agarose gels at roughly the same rate as double-stranded DNA fragments of 300 and 4000 bp, respectively (Fig. 2-7).

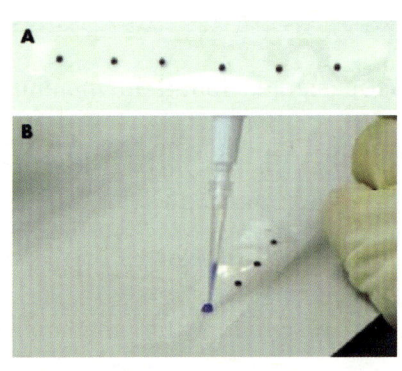

Fig. 2-6 Mixing the PCR samples with loading buffer

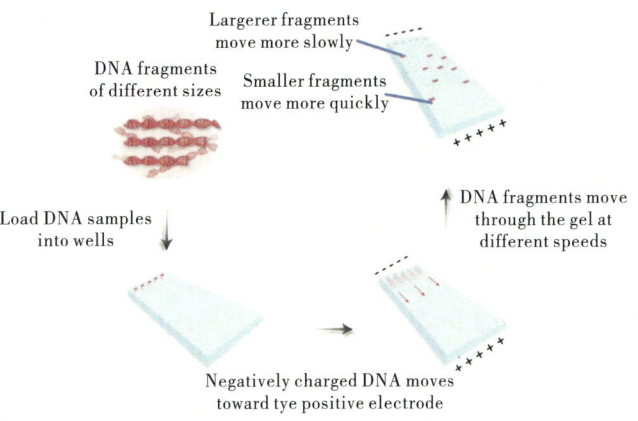

Fig. 2-7 DNA migration in agarose gel

4. Allelic frequency calculation

Once the genotypes of all students are collected (Fig. 2-8), the frequency of the PV92-Alu allele in the class population can be calculated.

First, determine the total number of possible alleles in the class population. Since each person has two alleles-one on each chromosome 16, the total number of people in the class that have a confirmed genotype will be by two alleles. Second, determine the total number of (+) alleles in the class population. Each homozygous positive student (+/+) has two (+) alleles, each homozygous negative (-/-) student has no (+) alleles, and each heterozygous students has one (+) allele. For example, if your class has 18 (+/+) students, 20 (+/-) students, and 10 (-/-) students, the number of (+) alleles in the class population would be $(18 \times 2) + (20 \times 1) + (10 \times 0) = 56$. Third, calculate the frequency of the (+) allele by dividing the number of (+) alleles in the class by the total number of possible alleles in the class population. In this example, the frequency would be $56/96 = 0.583$.

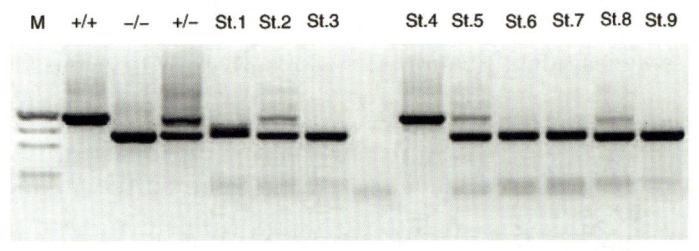

Fig. 2-8 PCR results of Alu analysis. M., marker; St., student

Section 2　Boy or Girl?

In mammals, male development and spermatogenesis is mainly governed by the action of a few genes located on the small and heterochromatic Y chromosome, notably by the testis-determining factor SRY protein, which is thought to trigger the pathway leading to a male phenotype. Birds have a reversed sex chromosome system compared with mammals in which females are heterogametic ZW, while males are homogametic ZZ (Fig. 2-9). Sex development in birds can therefore not rely on the same unique-to-male genetic signals found in mammals but additional ones. Indeed, no true SRY homolog has yet been detected in birds. Besides its reversed association to sex, the avian W chromosome resembles the mammalian Y chromosome in several ways. It is generally small and mostly heterochromatic, and in the chicken, it constitutes less than 2% of the total DNA content, with two families of satellite repeats comprising at least 75% of the chromosome.

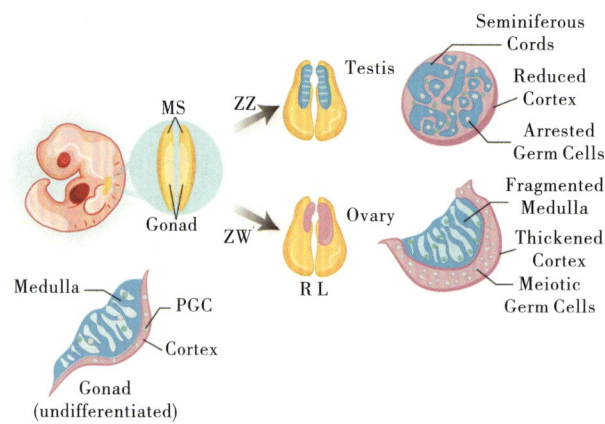

Fig. 2-9　The ZZ and ZW sex chromosome system of chickens

Accurate gender determination can be important along with species identification. Gender identification is relevant to the veterinarian, medical, and ecological science. The first gene on the W chromosome encodes chromodomain-helicase-DNA-binding protein 1 (CHD-1) with a possible globular role degree of conservation. Amplification of the CHD gene on the W and Z chromosomes can be used to determine the sex of most birds.

CHD-W is specific to females and CHD-Z is present in both sexes. Amplification of CHD should result in two fragments if the sample is from a female (W, Z) and only one if it is from a male (Z, Z). We can use the differences in the two fragments to distinguish males from females.

The chicken embryonic gonad is an important model to study animal organogenesis. In both sexes, the cyclin-dependent kinase 2-associated protein 1 (Cdk2ap1) is expressed in the head mesenchyme, rhombencephalon, otic vesicle, spinal neural tube, and forelimb bud of 4-day-old embryos. The expression level of Cdk2ap1 in males is significantly higher than that of females. In addition, in the genital ridge and hind limb bud of the 4-day-old chicken embryo, Cdk2ap1 is expressed in males. It is known that Cdk2ap1 is downregulated as embryos develop, and its expression in the female gonad is higher than that in the male in the embryo at 6.5-10 days old. The Cdk2ap1 expression differences between male and female embryos need further elucidation. It is suggested that Cdk2pa1 may play a role in the gonad development and sexual differentiation of the chicken embryo based on its differential expression.

Currently, there is no definite research about Cdk2ap1 in domestic and overseas chicken embryos, and the details pertaining to when, where, and how the gene works remain unclear.

Objectives

❖ Determining the sex of a chicken embryo

❖ Strengthening the necessary skills of PCR

❖ Strengthening the necessary skills of agarose gel electrophoresis

Materials

Reagents: PCR Kit (TaKaRa); Specific primers; DNase, RNase-free H_2O

Supplies: Pipette tips; PCR tubes (200 μL); Ice buckets; Marker pen

Equipment: Pipette; Microcentrifuge; Thermal cycler

Activity Protocol

1. Setting up PCR reactions

Method Reference: Page 18

When programming the PCR reaction, set the thermal cycler with the following program:

Initial denature: 95 ℃ for 5 min

35 cycles of: 95 ℃ for 30 s; 55 ℃ for 30 s; 72 ℃ for 30 s

Final extension: 72 ℃ for 2 min

Hold: 15 ℃

2. Preparing and running standard agarose gel

Method Reference: Page 19

3. Prospective PCR result

The result for the determination of embryos' gender is shown below (Fig. 2-10).

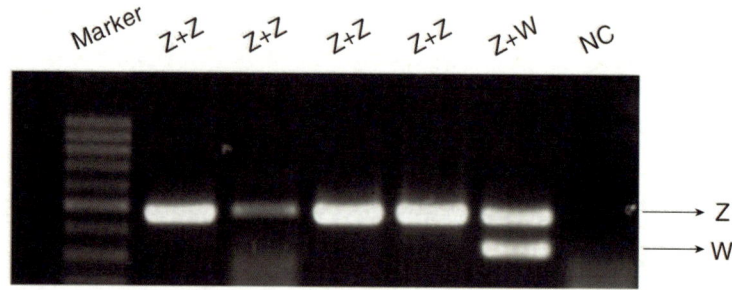

Fig. 2-10 The typical result for the sex determination of embryos

Section 3   Know More

1. Noninvasive preimplantation genetic testing for aneuploidy

Due to delayed childbearing age, social pressure, and bad lifestyle, some people have suffered from reproductive challenges. As a result, the assisted reproductive technology has been developed, which includes *in vitro* fertilization (IVF). IVF became a medical treatment for infertility in the 20th century, bringing hope to many couples who were unable to conceive naturally.

However, preimplantation genetic testing methods are essential for a successful pregnancy and a healthy fetus. Among them, the preimplantation chromosome aneuploidy test (PDT-A) refers to the use of a variety of techniques to assess whether the chromosome number and structure of early embryos are normal before embryo implantation. At present, the most commonly used chromosome aneuploidy detection is to extract cells from the embryo for trophectoderm biopsy. Although using this method to screen for euploid embryos can improve the success rates of embryos transferred, the accuracy, safety, and applicability of this method are always controversial.

The accuracy of the embryo trophoblast ectoderm (TE) biopsy is related to embryonic chimeras. Chimera refers to the presence of two or more different genotypes of cells in the same embryo. In embryos, up to 30%-40% of human blastocysts are chimeras, with euploidy and aneuploidy chimeras forming between 2.0%-2.9% and 14.0%-17.3% of the time, respectively, which inevitably leads to the failure of TE biopsy results in reflecting the euploidy of the inner cell mass and the resulting fetus. In addition, the safety of performing a trophectoderm biopsy is also of concern. Although there is no clear evidence that the trophectoderm biopsy is harmful to embryos, its safety remains unknown and is of concern because the removal of TE cells could potentially harm the embryo and even affect its ability to implant (Fig. 2-11).

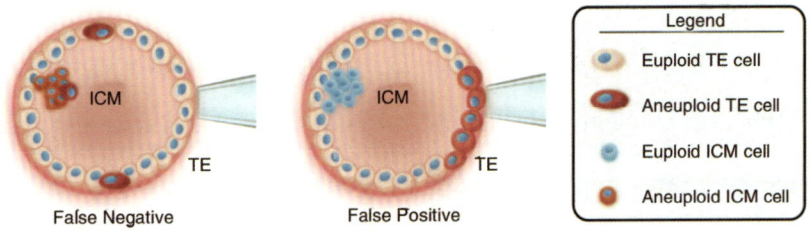

Fig. 2-11　False positives and false negatives arising from embryo mosaicism in TE biopsy PGT-A

On the basis of this limitation, noninvasive preimplantation genetic testing for aneuploidy (niPGT-A) was developed since 2016[2] and was further modified in 2019[3] to analyze embryonic DNA. The niPGT-A does not require TE cells to be extracted, but rather analyses the DNA that leaks from human blastocysts into the culture medium in which the cells are cultured (Fig. 2-12).

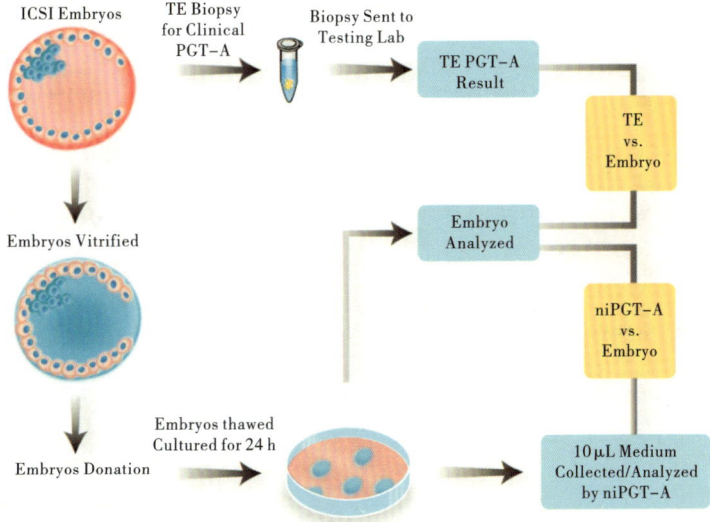

Fig. 2-12　Sample processing for the PGT-A analysis of the TE biopsy, embryo, and spent culture media

The researchers collected 52 frozen human blastocysts from donated samples that had previously been tested by TE biopsies. The blastocysts were then thawed and incubated separately in droplet cultures for 24 h. The cultures and embryos were collected, amplified, and sequenced. The detection of chromosomal aneuploidy was compared between niPGT-A and TE biopsy, which finally proved that the niPGT-A using free DNA in embryo culture medium was more accurate and reliable than using the TE biopsy.

niPGT-A, which has been shown to be more accurate than the existing methods, represents a new way of assessing embryos with lesser risk of chromosomally normal embryos being discarded.

2. The ASFV-A virus infects both domestic and wild swine

Pork, which accounts for more than one-third of the meat produced worldwide, is the most widely consumed meat in the world. Therefore, the emergence of a new virus, the African swine fever virus (ASFV), which causes a devastating and economically significant disease in both domestic and wild swine (Sus scrofa), is a critical strike to the pig industry.

ASFV is endemic in Africa, where it was first described in the early 1900s. In 1957, ASFV emerged outside of Africa in Portugal, and from 1960 to the early 20th century, the virus mainly spread across Europe to America. By the mid-1990s, ASFV had been eradicated in the Americas and Europe, with the exception of Sardinia, which has remained endemic since 1982. In 2007, ASFV was introduced to the Caucasus region and quickly spread into the Russian Federation and Eastern Europe, where it has continued to circulate. In August 2018, the ASFV outbreak in Asia was reported as the biggest animal disease outbreak to date, killing over 500 million pigs worldwide and putting thousands of farmers out of business.

ASFV belongs to the family Asfarviridae, which includes the single species ASFV, which has a linear dsDNA genome of 170-194 kb. The virus encodes 150-165 proteins, which have "essential" functions in virus replication and "non-essential" roles in host interactions, including the evasion of host defenses. For example, many host proteins are inhibited in the early innate immune responses, including the type I interferons and antiviral molecules involved in cell death pathways. Sequencing of the gene encoding the major capsid protein (B646L/p72) has revealed 23 different genotypes of ASFV.

Without effective treatment and vaccines, early and accurate diagnoses are effective alternative ways to rapidly control outbreaks in infected geographical regions, followed by movement restrictions and stamping-out policies. The current gold standard for ASFV diagnostics in the laboratory is based on traditional PCR, which can be used as a viral genome detection technique based on nucleic acid marker amplification. On the other hand, virus isolation based on viral replication in susceptible primary cells has also been used for ASFV diagnosis. Enzyme-linked immunosorbent assay (ELISA), a technique used to detect viral antigens in the host sample, has also been widely used for ASFV detection. However, a reliable, inexpensive, highly sensitive, and high-throughput ASFV detection method that is suitable for clinical diagnosis is still unexplored.

Recently, researchers present an advanced system for rapid and accurate detection of ASFV. Because ASFV is composed of a dsDNA genome, a clustered regulatory interspaced short palindromic repeats (CRISPR)/CRISPR-associated protein (Cas) 12a detection scheme was developed by cleaving ssDNA probes after ASFV in situ hybridization with a CRISPR RNA (crRNA)-programmed Cas12a. Furthermore, they demonstrated the high stability of the Cas12a system, extending the detection limit to 100 fM with longer incubation time. This compact detection system is automated, integrated, small, lightweight, inexpensive, and ready to be used for on-site ASFV detection or other DNA-based pathogen detection[4].

References

[1] BATZER M A, DEININGER P L, et al. Alu repeats and human genomic diversity[J]. Nat Rev Genet., 2002, 3: 370-379.

[2] XU J, FANG R, CHEN L, et al. Noninvasive chromosome screening of human embryos by genome sequencing of embryo culture medium for in vitro fertilization[J]. Proceedings of the national academy of sciences, 2016, 113: 11907-11912.

[3] LEI H, BERHAN B, et al. Noninvasive preimplantation genetic testing for aneuploidy in spent medium may be more reliable than trophectoderm biopsy[J]. Proceedings of the national academy of sciences, 2019, 116: 14105-14112.

[4] HE Q, YU D M, et al. High-throughput and all-solution phase African Swine Fever Virus (ASFV) detection using CRISPR-Cas12a and

fluorescence based point-of-care system[J]. Biosens bioelectron, 2020, 154: 112068. doi: 10.1016/j.bios.2020.112068.

Postlab Focus Questions

1. What techniques are PCR-based on?

2. How do you know whether the embryo is alive?

3. Is the punching position on the embryo important?

4. What is the difference in the sex development between plants and chicken embryos?

5. Could you find other ways to recognize the gender of the embryo?

(Written by Zhang Yan, Cao Yongchang)

Chapter 3   RNA Extraction and RT-PCR

Section 1   RNA Extraction and Purification

Background Reading

One of the first decisions that a researcher has to make when detecting or quantitating RNA is whether to isolate the total RNA or only the poly (A)-selected RNA (commonly referred to as mRNA). This choice is further complicated by the bewildering array of RNA isolation kits available in the marketplace. In addition, the downstream application intended influences this choice. The following section is a short summary to help in the decision process.

From an application point of view, total RNA might suffice for most applications, and it is frequently the starting material in applications ranging from the detection of an mRNA species by northern blotting to the quantitation of a message by RT-PCR. The preference for total RNA reflects the challenge of purifying enough poly (A) RNA for the application (mRNA comprises <5% of cellular RNA), the potential loss of a particular message species during poly (A) purification, and the difficulty in quantifying small amounts of purified poly (A) RNA. If the data generated with total RNA is not satisfied, using poly (A) RNA instead might provide the sensitivity and specificity required for the specific application. The obtained experimental data will provide the best guidance in deciding whether to use total or poly (A) RNA. Being flexible and open minded is required as there are many variables to consider when making this decision.

Poly (A) RNA is essential in two situations: cDNA library construction and the preparation of labeled cDNA for gene arrays. To avoid generating cDNA libraries with large numbers of ribosomal clones and non-specific labeled cDNA, it is crucial to start with Poly (A) RNA for these procedures.

Objectives

❖ Understanding the principles and significance of total RNA extraction

❖ Grasping the basic method for rapid preparation of total RNA of animal tissue

❖ Detection and quantitation of RNA

Materials

Tissue: Muscle of EGFP-transgenic pig

Reagents: TRIzol; Chloroform; Isopropanol; DEPC H_2O; 75% EtOH (prepared with DEPC H_2O)

Supplies: Pipette tips; Microcentrifuge tubes; Ice buckets; Marker pen; Floater

Equipment: Pipette; Microcentrifuge (with cooling system); Water bath; Spectrometer; Gel electrophoresis apparatus

General Safety Requirements

• Perform all work on ice unless otherwise indicated.

• Always wear clean gloves (do not touch your face/hair or "dirty" surfaces with your gloves and assume they remain nuclease-free).

• Use filter tips and other RNase-free consumables.

• Use diethyl pyrocarbonate (DEPC)-treated water at all times.

• Dispose of hazardous chemicals (Trizol and chloroform) in clearly labeled containers.

Activity Protocol

Part 1. RNA isolation

(1) Crush tissue samples (–0.6 g) using the chilled mortar and pestle. Occasionally add liquid nitrogen to the mortar to keep the sample submerged and cold. Grind the tissue sample until it becomes powdery. Allow liquid nitrogen inside the sample tube to evaporate. Divide the sample into two 1.5 mL RNase-free tubes and add 1 mL of Trizol to each tissue slurry. Mix thoroughly and vortex vigorously (Fig. 3-1).

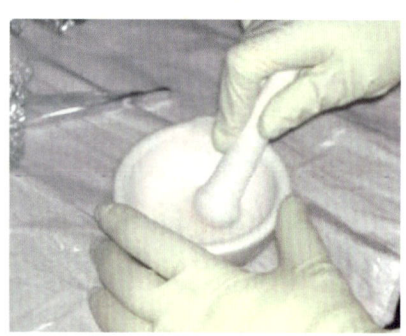

Fig. 3-1 Grinding tissue with the mortar and pestle

(2) Centrifuge at 12,000 × g at 4 ℃ for 5 min. Transfer the supernatant to a fresh RNase-free tube.

(3) Add 0.2 mL of chloroform/1 mL of Tri-

zol, cap the tube securely, and shake tube vigorously by hand for 15 s. Incubate at room temperature for 10 min or until the separation of layers is apparent. The mixture should separate into three layers: a lower red phenol-chloroform phase, an interphase (DNA, and carbohydrates, proteins, and other cellular debris), and a colorless upper aqueous phase (containing RNA, Fig. 3-2).

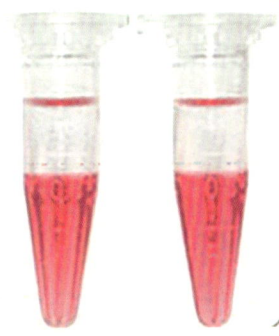

Fig. 3-2 Separation of water and organic phases after centrifugation

(4) Centrifuge at 12,000 × g at 4 ℃ for 15 min.

(5) Angle the tube at 45° and transfer 0.4 mL of the aqueous phase to a fresh RNase-free tube. Be very careful that only the aqueous phase is transferred. Sacrifice some sample rather than risk contamination.

(6) Add 0.4 mL of pre-chilled isopropanol/1 mL of TRIzol, mix well, and incubate at room temperature for 10 min.

(7) Centrifuge at 12,000 × g at 4 ℃ for 10 min.

(8) Remove the supernatant by pouring. The pellet is the total RNA.

(9) Add 1 mL of the 75% EtOH/1 mL of TRIzol used in the initial homogenization and vortex vigorously.

(10) Centrifuge at 8,000 × g at 4 ℃ for 10 min.

(11) Aspirate the supernatant and centrifuge at 12,000 × g at room temperature for 1 min. Use a micropipette to carefully withdraw any remaining EtOH. Be careful not to aspirate the pellet.

(12) Dry the pellet by air-drying for 20-30 min. Do not dry RNA by centrifugation under vacuum. Do not allow the RNA to dry completely because the pellet can lose solubility.

(13) Resuspend the pellet in 20-25 μL of DEPC H_2O by mixing the solution up and down several times through a pipette tip.

(14) Use the spectrophotometer to measure the concentration (260 nm) and ration (260 nm vs. 280 nm) of total RNA purified.

(15) Store the RNA at −80 ℃. Do not freeze and thaw frequently.

Part 2. RNA electrophoresis

(1) Wash the gel tank and combs with detergent, rinse and dry them with 70% EtOH or decontaminate them using the RNAseZap detergent. Use fresh 1 × TAE to

prepare and run the gel.

(2) Remove 1 μg of RNA for electrophoresis.

(3) Refer to the activity protocol in the DNA and PCR section.

Information Box 3-1: Nine ways to improve RNA isolation

Effective preparation of RNA is a fundamental technique that is required for a wide variety of exciting and information-rich analysis techniques, including next-generation sequencing, reverse transcription qPCR (RT-qPCR), northern blot analysis, and cRNA production. The fidelity of transcrip—tome representation and the quality and quantity of recovered RNA will significantly impact the resulting analysis.

(1) Immediate inactivation of endogenous, intracellular RNases.

(2) Proper tissue storage conditions.

(3) Thorough homogenization of samples.

(4) Pretreat homogenate before isolating RNA.

(5) Best RNA isolation method.

(6) DNase treatment.

(7) Reduce exposure to environmental RNase.

(8) Proper precipitation.

(9) RNA storage under −80 ℃.

Section 2   RT-PCR

Background Reading

RT-PCR is the most sensitive method for detecting and quantifying mRNA. Theoretically, even extremely low-abundance messages can be detected with this technique. The total RNA is routinely used as the template for RT-PCR, but some cloning situations and rare messages require the use of poly (A) RNA.

Reverse transcription may be primed with either Oligo $(dT)_{15}$ or random primers or specific primers. Choosing Oligo $(dT)_{15}$ when priming at the 3′ poly (A) region is desired, while choosing random primers when priming throughout the length of the RNA is desired. Oligo $(dT)_{15}$ is frequently used when cDNA will be used for cloning and RT-PCR. Random primers are sometimes preferred for cDNA used in RT-PCR, especially when the PCR primers target a region near the 5′-end of the RNA. The target sequence should be known when using specific primers (Fig. 3-3).

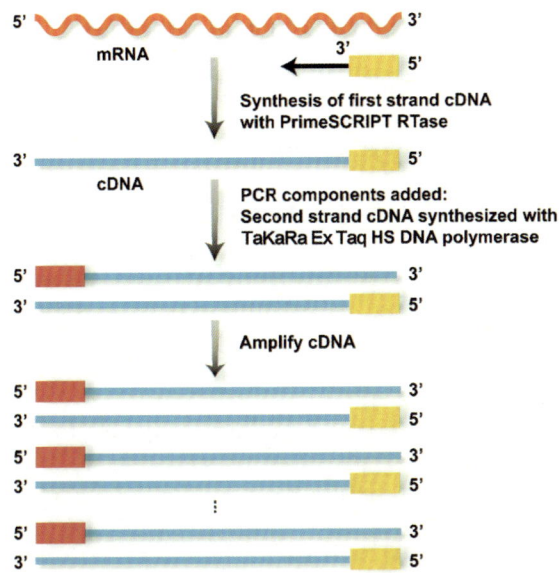

Fig. 3-3 The mechanism of RT-PCR

Note: *A school of thought concerning RT-PCR considers it advisable to treat the sample RNA with DNase I, since no purification method produces RNA completely free of residual genomic DNA contamination. RT-PCR is sensitive enough that even small amounts of genomic DNA contamination can cause false positives.*

Objectives

❖ Learning the technique of RT-PCR

❖ Detection of EGFP in animal tissues

Materials

Reagents: PrimeScript RT-PCR Kit (Takara); Primers for the target gene and the house-keeping gene

Supplies: Pipette tips; Microcentrifuge tubes; Ice buckets; Marker pen

Equipment: Thermal Cycler for PCR; Pipette; Microcentrifuge; Electrophoresis apparatus

Activity Protocol

Part 1. Reverse transcription reaction

(1) Prepare a 10 μL reaction by adding the following reagents in the order listed (Table 3-1):

Table 3-1 Reaction for denaturing and annealing

Component	Amount
dNTP Mixture (10 mM)	1 μL
Oligo $(dT)_{15}$ Primer (2.5 μM)	1 μL
Total RNA	1 μg
RNase Free dH_2O	up to 10 μL

(2) Proceed to the denaturing and annealing PCR stages: at 65 ℃ for 5 min, and at 4 ℃.

(3) Briefly spin the reaction mixture and prepare the following reverse transcription reaction (Table 3-2).

Table 3-2 Reverse transcription reaction

Component	Amount (μL)
Above reaction mixture	10
5 × PrimeScript Buffer	4
RNase Inhibitor (40 U/μL)	0.5
PrimeScript RTase	0.5
RNase Free dH_2O	5
Total Volume	20

(4) Reverse transcription reaction is performed as follows: 42 ℃ for 30 min, 95 ℃ for 5 min while on ice.

Part 2. PCR amplification

(1) Prepare a 20 μL PCR amplification mix by combining the following reagents (Table 3-3).

Table 3-3 PCR reaction

Component	EGFP	GAPDH
10 × PCR Buffer II	2 μL	2 μL
dNTP Mixture (10 mM)	0.8 μL	0.8 μL
EGFP-F Primer (10 μM)	0.5 μL	
EGFP-R Primer (10 μM)	0.5 μL	
GAPDH-F Primer (10 μM)		0.5 μL
GAPDH-R Primer (10 μM)		0.5 μL
Takara Ex Taq HS (5 u/μL)	0.2 μL	0.2 μL
First-strand cDNA reaction	1 μL	1 μL
RNase Free dH$_2$O	15 μL	15 μL

(2) Thermal cycling is performed, and the profiles included one cycle of 5 min at 95 ℃, 35 cycles of 30 s at 94 ℃, 30 s at 60 ℃, 30 s at 72 ℃, and one cycle of 5 min at 72 ℃.

Information Box 3-3: Glyceraldehyde 3-phosphate dehydrogenase (GAPDH) is an energy metabolism-related enzyme in the glycolytic pathway. GAPDH is often stably and constitutively expressed at high levels in most tissues and cells, it is considered as a housekeeping gene. For this reason, GAPDH is commonly used as a loading control for RT-PCR.

Part 3. DNA electrophoresis

Refer to the gel electrophoresis protocol described in the DNA and PCR sections.

Section 3 TA Cloning

Objective

- Ligation of the cloned cDNA fragment into the TA vector

Materials

Reagents: Axygen® AxyPrep™ DNA Gel Extraction Kit and pGEM-T Easy ligation kit (Promega)

Supplies: Pipette tips; Microcentrifuge tubes; Ice buckets; Marker pen; Floater

Equipment: Pipette; Microcentrifuge (with cooling system); Water bath; Spectrometer; Gel electrophoresis apparatus; Sharp scalpel; UV protective glasses and UV transilluminator

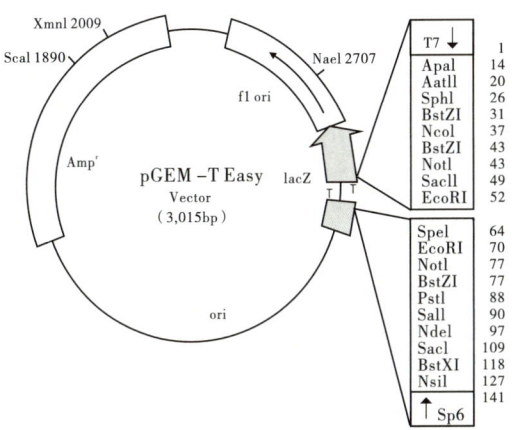

Fig. 3-5　The map of the pGEM®-T Easy Vector

⚙ Activity Protocol

Part 1. Gel extraction

(1) Excise the agarose gel slice containing the DNA fragment of interest with a clean, sharp scalpel under ultraviolet illumination. Briefly place the excised gel slice on an absorbent towel to remove any residual buffer. Transfer the gel slice into a container piece, plastic wrap, or weighing boat. Mince the gel into small pieces and weigh its content. In this application, the weight of the gel is considered equivalent to its volume. For example, 100 mg of gel is equivalent to 100 μL of gel volume. Transfer the gel slice into a 1.5 mL microcentrifuge tube.

Note: Alternatively, the gel slice can be placed into the 1.5 mL microcentrifuge tube and then crushed with a pipette tip or other suitable device. Spin the tube for 30 s at 12,000 × g to consolidate the gel at the bottom of the tube. Use graduations in pipetting volume to estimate the volume of the agarose gel.

(2) Add a 3 × sample volume of Buffer DE-A.

Note: The color of Buffer DE-A is red. This color is used to add contrast in the next step so that any pieces of insolubilized agarose can be visualized.

(3) Resuspend the gel in Buffer DE-A by vortexing. Heat at 75 ℃ until the gel is completely dissolved (typically lasts 6-8 min). Heat at 40 ℃ if a low-melting agarose gel is used. Intermittent vortexing (every 2-3 min) will accelerate gel solubilization.

Note: The gel must be completely dissolved, otherwise the DNA fragment recovery will be reduced. Do not heat the gel for longer than 10 min.

(4) Add 0.5 × Buffer DE-A volume of Buffer DE-B into the sample and thoroughly mix. If the DNA fragment is less than 400 bp, supplement further with a 1 × sample volume of isopropanol.

Example: For a 1% gel slice equivalent to 100 μL, add the following:

- 300 μL Buffer DE-A
- 150 μL Buffer DE-B

If the DNA fragment is < 400 bp, you would also add:

- 100 μL of isopropanol

Note: *The color of the mixture will turn yellow after the addition of Buffer DE-B. Please make sure the contents are a uniform yellow color before proceeding.*

(5) Place a Spin Miniprep column into a 2 mL microfuge tube (provided). Transfer the solubilized agarose from Step 4 into the column. Centrifuge at 12,000 × g for 1 min.

(6) Discard the filtrate from the 2 mL micro-centrifuge tube. Place the Miniprep column back into the 2 mL microcentrifuge tube and add 500 μL of Buffer W1. Centrifuge at 12,000 × g for 30 s.

(7) Discard the filtrate from the 2 mL microcentrifuge tube. Place the Miniprep column back into the 2 mL microcentrifuge tube and add 700 μL of Buffer W2. Centrifuge at 12,000 × g for 30 s.

Note: *Make sure that 95%-100% ethanol has been added into the Buffer W2 concentrate. Make a notation on the bottle label for future reference.*

(8) Discard the filtrate from the 2 mL microcentrifuge tube. Place the Miniprep column back into the 2 mL microcentrifuge tube. Add a second 700 μL aliquot of Buffer W2 and centrifuge at 12,000 × g for 1 min.

Note: *Two washes with Buffer W2 are used to ensure the complete removal of salt, eliminating potential problems in subsequent enzymatic reactions, such as ligation and sequencing reactions.*

(9) Discard the filtrate from the 2 mL microcentrifuge tube. Place the Miniprep column back into the 2 mL microcentrifuge tube. Centrifuge at 12,000 × g for 1 min.

(10) Transfer the Miniprep column into a clean 1.5 mL microcentrifuge tube (provided). To elute the DNA, add 25-30 μL of Eluent or deionized water to the center of the membrane (without rupturing the membrane). Let it stand for 1 min at room temperature. Centrifuge at 12,000 × g for 1 min (Fig. 3-4).

Note: *Pre-warming the Eluent at 65 ℃ will generally improve elution efficiency.*

Deionized water can also be used to elute the DNA fragments.

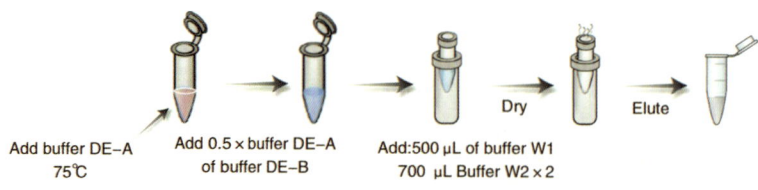

Fig. 3-4 Summary of the gel extraction procedure

Part 2. TA ligation

(1) Measure the concentration (260 nm) of the purified PCR product.

(2) The pGEM-T Easy Vectors (Fig. 3-5) are approximately 3 kb long and are supplied at 50 ng/μL. To calculate the appropriate amount of PCR product (insert) to include in the ligation reaction, use the following equation:

ng of insert = ng of vector × kb size of insert × 3 (insert: vector molar ratio)/ kb size of the vector.

(3) Set up ligation reactions as described below (Table 3-4). Vortex the 2 × Rapid Ligation Buffer vigorously before each use.

Use 0.5 mL microcentrifuge tubes, which are known to have low DNA binding capacity.

Table 3-4　Ligation reaction

Component	Amount (μL)
2 × Rapid ligation buffer	5
pGEM-T Easy vector (50 ng)	1
PCR product	X
T4 DNA ligase (3 Weiss units/μL)	1
ddH$_2$O to a final volume	10

(4) Mix the reactions by pipetting up and down. Incubate the reactions 1 h at room temperature. Alternatively, incubate the reactions overnight at 4℃ to obtain the maximum number of transformants.

Postlab Focus Questions

1. Is it possible to predict the total RNA yield from a certain mass of tissue?

2. Is there protein in your RNA preparation, and if so, should it be concerned?

3. Is your RNA physically intact? Does it matter?

4. How does DEPC inhibit RNase? Is more DEPC better?

5. What protocol modifications should be implemented to isolate RNA from difficult tissues, such as fibrous tissue, lipid-and polysaccharide-rich tissue, nucleic acid and nuclease-rich tissue, and hard tissue, as well as from bacteria and yeast?

(Written by Huang Shengfeng)

Chapter 4 Bacterial Transformation and Plasmid Purification

Section 1 Bacterial Transformation and Culture

Background Reading

Bacteria Transformation

Now that the recombinant DNA is ligated into a suitable vector, such as a plasmid or bacteriophage, it is important to introduce it into an organism able to amplify it, such as bacteria. Under normal conditions, bacteria are unable to take up these free plasmids and must undergo a transformation process in order to become "competent" to do so. Treating bacteria with a solution of calcium chloride followed by "heat shocking" enables the transformation. The plasmid and bacterial solutions are mixed, and some of the bacteria will take up the plasmids. Less than one percent of bacteria are transformed; each bacterium should take up a maximum of one plasmid and should only contain one recombinant molecule. By using a plasmid containing a gene that codes for resistance to an antibiotics, such as amp^R(which gives resistance to ampicillin), it can be determined which bacteria have taken up the plasmid. The bacteria are plated out at low concentration on agar plates containing the said antibiotics, and if none had resistance before the treatment, only the ones containing the plasmid and its resistance gene will survive. As each bacterium forms a colony, each cell in the colony will contain a single type of DNA molecule.

Blue/White Screening

Another method to determine if the bacterium contains the plasmid of interest and that it is ligated with the DNA fragment in the right direction is by blue/white screening.

Alpha-complementation occurs when two inactive fragments of *E. coli* β-galactosidase associate to form a functional enzyme. Many of the plasmid vectors carry a short segment of *E. coli* DNA containing the regulatory sequences and the coding information for the first 146 amino acids of the lacZ β-galactosidase gene. Embedded in the coding region is a polyclonal site that maintains the reading frame and results in the harmless interpolation of a small number of amino acids into the amino-terminal fragment of β-galactosidase. Vectors of this type are used in host cells that express the carboxy-terminal portion of β-galactosidase. Although neither the host-encoded nor the plasmid-encoded fragments of β-galactosidase are independently active, they can associate to form an enzymatically active protein. This type of complementation, in which deletion mutants of the operator-proximal segment of the lacZ gene are complemented by β-galactosidase-negative mutants that have the operator-proximal region intact, is called α-complementation. Lac$^+$ bacteria that result from α-complementation are easily recognized because they form blue colonies in the presence of the chromogenic X-gal. However, the insertion of a foreign DNA fragment into the polyclonal site of the plasmid almost invariably results in the production of an amino-terminal fragment that is no longer capable of α-complementation. Therefore, bacteria carrying recombinant plasmids form white colonies. The development of this simple color test has vastly simplified the identification of recombinants constructed in plasmid vectors (Fig. 4-1).

Strains of bacteria commonly used for α-complementation do not synthesize significant quantities of lac repressors. If necessary, the synthesis of both fragments can be fully induced by IPTG, a nonfermentable lactose analog that inactivates the lacZ repressor (Fig. 4-1).

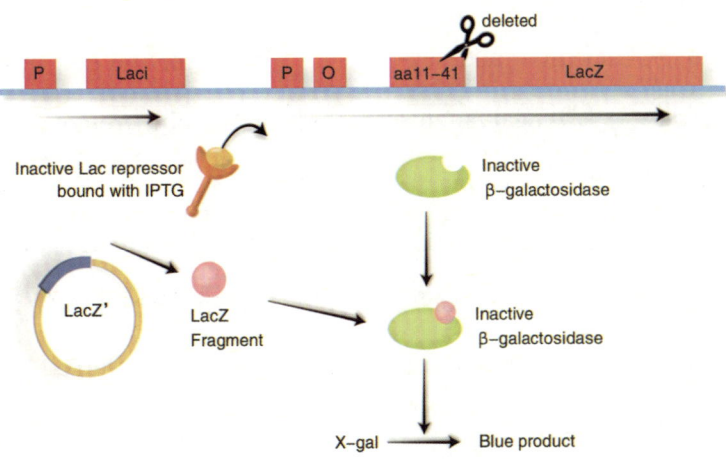

Fig. 4-1 A schematic representation of the blue/white assay used to screen for recombinant vectors

Objective

❖ Introduction of the recombinant plasmids into bacteria

Materials

Reagents: LB agar plates; X-Gal (20 mg/mL, dissolved in DMSO (dimethyl sulfoxide) or dimethyl formamide, sterilized by filtering, kept at −20 ℃); IPTG (24 mg/mL, water solution, sterilized by filtering, kept at 4 ℃); Glass rod spreader

Supplies: Pipette tips; Microcentrifuge tubes; Ice buckets; Marker pen; Floater

Equipment: Pipetter; Microcentrifuge; Water bath; Incubator; Clean bench

Active Protocol

(1) The competent cells JM109 (or DH5α) were used. The specification is: 100 μL competent cells × 10 tubes, 10 μL control vector (pUC19, 0.1 ng/μL) × 1 tube, 1 mL SOC broth × 10 tubes, stored at −80 ℃.

(2) Warm up your SOC medium and seven X-gal + IPTG + Amp + LB agar plates using a 37 ℃ incubator. Plates should be independently labeled as A, B, C, D, E, "positive control," and "negative control". Excessive moisture in the plates should be evaporated by heating before use.

(3) Prepare the water bath (42 ℃).

(4) Prepare seven 1.5 mL tubes (better with nuclease-free tubes) and label them as A, B, C, D, E, "positive control" and "negative control," and then chill them on ice (better on ice-water mixture).

(5) Thaw 2-3 tubes of competent cells on ice (not at room temperature!), gently homogenize the cells using a 200 μL pipette. Then add the cells to each pre-chilled 1.5 mL tube with 40-45 μL for A, B, C, D, and E, and 15 μL and 10 μL for the positive control and negative control, respectively.

(6) Thaw your ligation sample and add 3-4 μL of ligation sample and 1-2 μL of control vector directly into the corresponding tubes. Incubate for 30 min on ice.

(7) Using the water bath, heat-shock the mixture for 50-55 s at exact 42 ℃, and immediately transfer into ice for 5 min.

(8) Add 0.3-0.4 mL of SOC and incubate the tubes for 60 min in a 37 ℃ incubator.

(9) After the cell mixtures have incubated for 60 min, remove them from the incubator.

(10) Optional: add 20 μL X-Gal (20 mg/mL) and 20 μL IPTG (24 mg/mL) to the cells, and gently mix.

(11) Optional: gently pellet the cells at 2000 rpm for 3-5 min and re-suspend them in 100 μL fresh SOC or LB medium.

(12) Plate the transformed cells onto LB plates with a glass rod spreader. Place the plates in the 37 ℃ incubator. Keep plates upside down at 37 ℃ for 12-16 h.

(13) Take out the plate and observe the bacterial colonies. Calculate the ratios of white and blue colonies.

Information Box 4-1: Electroporation

This physical method was originally developed for eukaryotic cells and later adapted for bacteria. Thus far, it is the most efficient and reproducible way (after optimization) for the introduction of foreign DNA material into bacteria (10^{10} transform-ants/μg plasmid DNA and 80% of cells transformed). It is suitable for lineage DNA, big plasmids (e.g., BAC), double transformation, transformation-refractory cells, and scarce materials.

Information Box 4-2: Methods for competent bacterial cell preparation

The classical CaCl$_2$ method. Competent bacterial cells can be obtained by washing the collected cells at the logarithmic growth stage (OD600 = 0.35-0.4) using cocktails of solutions with CaCl$_2$. This simple, robust, and repeatable method yields a transformation efficiency of 10^5-10^7 colonies/μg supercoiled plasmid DNA.

The Inoue method. An efficiency of 10^8 colonies/μg plasmid DNA can be achieved for *E. coli*. Cell culture should be carried under 18 ℃.

The Hanahan method. It was developed by Douglas Hanahan in the 1970s (Fig. 4-2). High-efficiency transformation (5×10^8 colonies/μg plasmid DNA) is achieved by washing bacteria with TFB and FSB solutions, the so-called "liquid gold" in molecular cloning.

Fig. 4-2　Professor Douglas Hanahan

Section 2 Colony PCR Screening for the Reconstructed Plasmids

Background Reading

Colony PCR is a method used for screening bacterial (*E. coli*) colonies to validate the correct ligation or plasmid products. Selected colonies of bacteria are picked with a sterile toothpick or pipette tip from a growth agar plate and inserted into the PCR master mix. PCR is then conducted to determine if the colony contains the DNA fragment or plasmid of interest. Primers are used, which generate a PCR product of known size. Thus, any colonies which give rise to an amplified product of the expected size are likely to contain the correct DNA sequence.

Objective

❖ Identification of reconstructed plasmids with the cDNA fragment insert

Materials

Reagents: Taq system: Taq buffer (10×); dNTP (each 2.5 mM, totally 10 mM, kept at −20 ℃; 5′ and 3′ vector flanking primers (T7 and SP6) for the T-easy vector (10 μM for each)

Supplies: Tooth picks; Pipette tips; Microcentrifuge tubes; 0.2 mL PCR tubes or 96-well plate; Marker pens; Ruler

Equipment: Pipetter; Microcentrifuge; Thermocycler

Active Protocol

(1) Warm up the X-gal + IPTG + Amp + LB agar plates with a 37 ℃ incubator.

(2) Set up a 10 × PCR pre-mix as follows and distribute into 18-20 PCR tubes (200 μL) (Table 4-1).

Table 4-1 PCR reactions

In each tube	× 1 tube (μL)	× 20 tubes (μL)
ddH$_2$O	14.5	290
10 × PCR buffer	2.0	40
Forward Primer (10 μM)	1.0	20
Reverse Primer (10 μM)	1.0	20
dNTPs (10 mM)	1.5	30
Taqase	0.1	2.0
Total	20.0	400

(3) Add 20 μL of pre-mix to each PCR tube on ice.

(4) Touch a fresh toothpick onto the middle of a white-colored colony, dip it into a PCR tube, then streak it onto a fresh replicate agar plate marked with grids and numbers at the plate bottom of the plate (that is,

all selected colonies on a single agar plate).

(5) Repeat this for eight times or more, meaning that you pick 8-18 colonies.

(6) Save at least one tube for negative control (no DNA template is added).

(7) Save at least one tube for positive control. Use a colony that will definitely yield a product with your primers. If you do not have a positive colony, then you may use a small amount of plasmid DNA or genomic DNA from the blue-colored colony.

(8) Place the replicate agar plate in the incubator. Keep plates upside down at 37 ℃ for 12-16 h or overnight.

(9) Run the PCR program. While the PCR is running, prepare the agarose gel for it to be ready to analyze the PCR products.

PCR Program

Thermal cycling is performed, and profiles include one cycle of 5 min at 94 ℃, 35 cycles of 25 s at 94 ℃, 25 s at 55 ℃, and 20 s at 72 ℃, and another cycle of 2 min at 72 ℃, stopping at room temperature (16-25 ℃).

* Annealing temperature may need to be modified depending on your amplicon. The temperature at 55 ℃ works well for T7 and SP6.

Section 3   Purification of Plasmid DNA

Background Reading

Alkaline lysis, in combination with the detergent SDS, has been used for more than 20 years to isolate plasmid DNA from *E. coli*. Exposure of bacterial suspensions to the strongly anionic detergent at high pH opens the cell wall, denatures chromosomal DNA and proteins, and releases plasmid DNA into the supernatant. Although the alkaline solution completely disrupts base pairing, the strands of closed circular plasmid DNA are unable to separate from each other because they are topologically intertwined. As long as the intensity and duration of OH-exposure is not too high, the two strands of plasmid DNA combine once again when the pH is returned to neutral.

During lysis, the bacterial proteins, ruptured cell walls, and denatured chromosomal DNA becomes enmeshed in large complexes that are coated with dodecyl sulfate. These complexes are efficiently precipitated from solution when sodium ions are replaced by potassium ions. After the denat-

ured material has been removed by centrifugation, native plasmid DNA can be recovered from the supernatant. Alkaline lysis in the presence of SDS is a flexible technique that works well with all strains of *E. coli* and with bacterial cultures ranging in volume from 1 mL to >500 mL. The closed circular plasmid DNA recovered from the lysate can be purified in different ways and to a different extent according to the needs of the experiment.

Most widely used kits for plasmid purification are based on spin-column chromatography. Briefly, the bacteria harboring the plasmid is pelleted and lysed, and the released plasmid DNA is bound to a minicolumn in the presence of Binding Solution. Under these conditions, only DNA will bind to the column while most of the contaminating RNA and cellular proteinaceous components are removed in the flow-through. The bound DNA is then washed to remove any remaining impurities. Lastly, the purified DNA is eluted into 50-100 μL of Elution Buffer or water.

Objectives

- Amplification of the desired plasmid through bacterial cell culture
- Purification of a large number of plasmids from cultured bacteria

Materials

Reagents: LB medium with Ampicillin (tryptone, yeast extract, NaCl, 1N NaOH, and Ampicillin); 80% Glycerol; Omega Plasmid Mini kit; ddH$_2$O

Supplies: Tooth picks; Pipette tips; Microcentrifuge tubes; Marker pens

Equipment: Pipetter; Microcentrifuge; UV Spectrophotometer

Activity Protocol

Part 1. Preparation of the replicate agar plate

Retrieve the replicate agar plate from the incubator and store at 4 ℃ for later use.

Part 2. Preparation of LB (Luria-Bertani) medium (250 mL) for cell culture

(1) Dissolve 2.5 g tryptone, 1.25 g yeast extract, and 2.5 g NaCl in 220 mL deionized water.

(2) Adjust the pH of the medium to 7.0 using 1-5 N NaOH and bring the volume up to 250 mL.

(3) Autoclave on a liquid cycle for 20 min at 15 psi.

(4) Allow the solution to cool down to 55 ℃ and add Ampicillin to a final concentration of 50 μg/mL.

(5) Store at room temperature or 4 ℃.

(6) Prepare five 7 mL or 15 mL tubes,

one for the negative control (no bacterium inoculation) and one for each bacteria culture from each student (e.g., 3 for white colonies and 1 for the blue colony).

(7) Add 3 mL LB medium to each tube.

(8) Pick bacteria colonies from your replicate agar plate and insert into the tube.

(9) Seal the tube loosely with caps, and place them in a 37 ℃ rocking incubator (200-300 rpm) overnight.

Part 3. Retract the cell cultures from the rocking incubator

Store them at 4 ℃ for a long period (>1 day) or room temperature for a short period (within 1 day) for later use.

Part 4. Plasmid purification using a commercial kit

(1) OPTIONAL: Prepare and equilibrize 4-6 filter columns with 100 μL 3 M NaOH, centrifuge at 12,000 × g for 30-60 s, then decant the flow-through liquid.

(2) Pellet 1.5-4.5 mL bacteria by centrifugation at 12,000 × g for 1 min at room temperature. Decant or aspirate medium and discard. Re-suspend the bacterial pellet by adding 250 μL of Solution I (with supplement of RNase A) and vortex (or pipet up and down) to mix thoroughly. Complete resuspension (no visible cell clumps) of cell pellet is vital for obtaining good yields.

(3) Add 250 μL of Solution II and gently mix by inverting and rotating the tube several times to obtain a clear lysate. A 2-3 min incubation may be necessary. Avoid vigorous mixing as this will shear chromosomal DNA and lower plasmid purity. Do not allow the lysis reaction to proceed more than 5 min. (Store Solution II tightly capped when not in use to avoid acidification from CO_2 in the air.)

(4) Add 350 μL of Solution III and mix immediately by inverting the tube several times until a flocculent white precipitate forms. It is vital that the solution is mixed thoroughly and immediately after the addition of Solution III to avoid localized precipitation.

(5) Centrifuge at 12,000 × g for 10 min at room temperature. A compact white pellet will form. Promptly proceed to the next step.

(6) Add the cleared supernatant by carefully aspirating it into a clean filter column assembled in a provided 2 mL collection tube. Ensure that the pellet is not disturbed and that no cellular debris has been carried over into the filter column. Centrifuge for 1 min at 12,000 × g at room temperature to completely pass lysate through the column.

(7) OPTIONAL: Discard flow-through liquid and reuse the 2 mL collection tube. Add 500 μL of HBC Buffer to wash the col-

umn. Centrifuge for 1 min at 12,000 × g at room temperature to completely pass solution through the column. Step ensures that residual protein contaminations are removed, thus ensuring high quality DNA that will be suitable for downstream applications. HBC Buffer must be diluted with isopropanol before use.

(8) Discard flow-through liquid and reuse the 2 mL collection tube. Add 600 μL of DNA Wash Buffer diluted with absolute ethanol to wash the column. Centrifuge for 1 min at 12,000 × g at room temperature to completely pass solution through the column and discard flow-through liquid.

(9) Optional Step: Repeat the wash step with another 600 μL of DNA Wash Buffer diluted with absolute ethanol.

(10) Centrifuge the empty column for 2 min at 12,000 × g to dry the column matrix. Do not skip this step, it is critical for good yields.

(11) Place the column into a clean 1.5 mL microcentrifuge tube. Add 30 μL to 50 μL (depending on desired concentration of final product) of Elution Buffer (10 mM Tris-HCl, pH 8.5) or sterile deionized water directly onto the column matrix and let it sit at room temperature for 1-2 min. Centrifuge for 1 min at 12,000 × g to elute DNA. An optional second elution will yield any residual DNA, though at a lower concentration (Fig. 4-3).

(12) Yield and quality of DNA: Determine the absorbance of an appropriate dilution of the sample at 260 nm and then at 280 nm.

Note: *A ratio of (A 260)/(A 280) is an indication of nucleic acid purity. A value greater than 1.8 indicates > 90% nucleic acid purity. Alternatively, yield (as well as quality) can sometimes be best determined by agarose gel electrophoresis by comparison to DNA samples of known concentrations. Typically, the majority of the eluted DNA is in monomeric supercoil form, although concatamers may also be seen.*

(13) Perform gel electrophoresis to analyze the plasmid as described before.

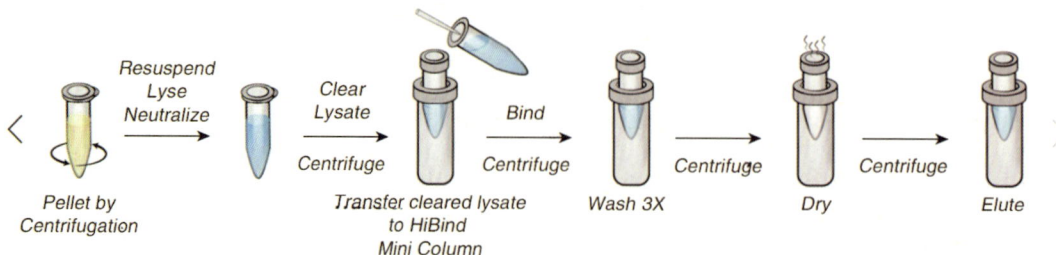

Fig. 4-3 Workflow of plasmid purification

Section 4　Know More

1. RNA sequencing

Instead of cloning the required RNA using RT-PCR with specific primers, a current popular procedure is to sequence the whole RNA library by using high-throughput sequencing technology, which allows the determination of millions or hundreds of millions of mRNA at low cost. This is called RNA sequencing (RNA-seq), which means capturing the total RNA from the collection of cells, and then sequencing it using high-throughput sequencing machines in order to determine which genes are active or expressed in those cells. RNA-seq was developed more than a decade ago and since then has become a popular tool in molecular biology that is shaping nearly every aspect of our understanding of genomic function.

Basic RNA-seq experiments include RNA extraction, fragment RNA, cDNA synthesis, adaptor ligation, PCR amplification, sequencing, and data analysis. There are three main RNA-seq technologies:

- Short-read cDNA (Illumina and Ion Torrent platform)
- Long-read cDNA (PacBio and ONT platform)
- Long-read RNA (ONT platform)

Comparison of the three sequencing platforms:

(1) Illumina.

After library preparation, individual cDNA molecules are clustered on a flowcell for sequencing by synthesis using blocked fluorescently labeled nucleotides. In each round of sequencing, the growing DNA strand is imaged to detect which of the four fluorophores has been incorporated. Reads of 50-500 bp can be generated[1].

(2) Pacific Biosciences (PacBio).

After library preparation, individual molecules are loaded into a sequencing chip, where they bind to a polymerase immobilized at the bottom of a nanowell. As each of the fluorescently labeled nucleotides is incorporated into the growing strand, they fluoresce and are detected. Reads of up to 50 kb can be generated[2].

(3) Oxford Nanopore (ONT).

After library preparation, individual molecules are loaded into a flowcell, where motor proteins, which are attached during the adapter ligation reaction, dock with nanopores. The motor protein controls the translocation of the RNA strand through the nanopore, causing a change in current that is processed to generate sequencing reads of 1-10 kb[3].

Applications of RNA-seq:
- Differential gene expression
- Differential splicing of mRNAs
- Transcriptional and translational dynamics
- RNA structure
- RNA-RNA and RNA-protein interactions
- Single-cell RNA-seq
- Spatialomics

2. Homologous recombination in cloning technology

Homologous recombination (HR) is a type of genetic recombination that happens in many species (e.g., eukaryotes, bacteria, and viruses) and depends on homologous sequences identification. Recently, some studies exploited its machinery to develop recombinational cloning technology. On this basis, many companies optimized the HR reaction system and produced commensal HR-based cloning kits working *in vitro* efficiently.

HR-based cloning technology reduces vector self-ligation significantly. In addition, it also enables precise directional insertion of fragments on vectors with only one single cloning site, which reduces the limitation caused by the lack of suitable cloning sites (Fig. 4-4).

General HR-based cloning technology contains four steps:

(1) Prepare linearized vectors by digesting circle vectors with restriction enzymes or by RT-PCR.

(2) Amplify insertion fragments with optimal PCR primers, which contain additional 15-25 bp homologous sequences of linearized vectors and gene-specific amplification sequences. To make PCR fragments and linearized vectors, identify each other by overlapping ends.

(3) Mix PCR products with linearized vectors in a certain molar ratio (according to different enzymes, the general ratio is 2∶1, less than the required by most traditional ligation methods). Add the recombination enzyme mixture and allow the reaction to happen at the optimum temperature.

(4) Transform competent bacterial cells with the recombinant plasmids following the general transformation methods.

The entire procedure is faster and easier than the traditional T4 ligation method. Multiple fragments can even be assembled in one single reaction.

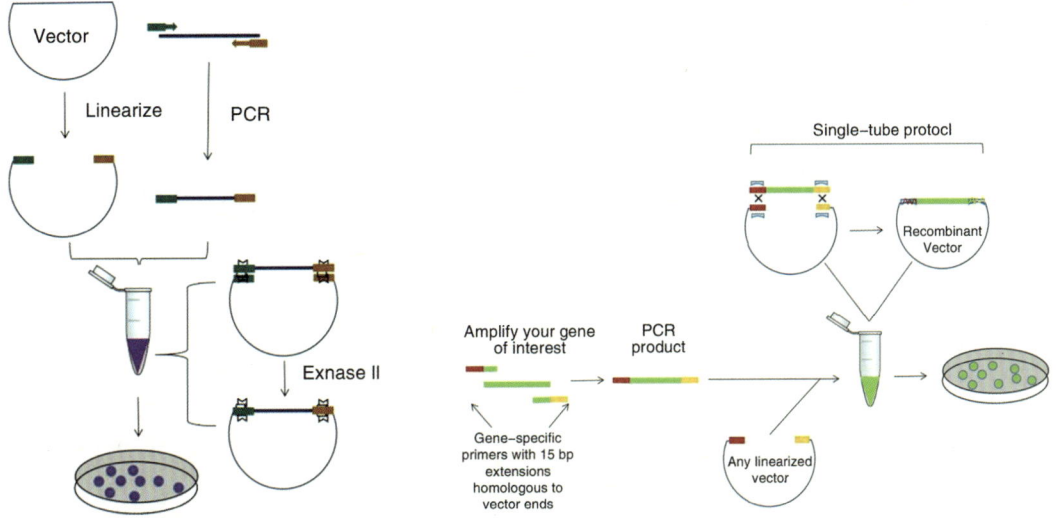

Fig. 4-4 General HR-based cloning technology

≫References

[1] https://www.illumina.com.

[2] https://www.pacb.com.

[3] https://nanoporetech.com/.

Postlab Focus Questions

1. Why dry the plates before dispensing? Why re-suspend the transformed cells?

2. Why keep the plates upside down when growing the transformed bacteria?

3. What is the length of your PCR product?

4. Why are extra procedures not needed to lyse the bacteria in the PCR screening?

5. What is the advantage and disadvantage of using T7/SP6 primers or specific primers?

6. Why is the bacteria culture tube sealed loosely?

7. Why handle the inverting action gently after adding solutions 1, 2, and 3? What would happen after drastically shaking the tube?

(Written by Huang Shengfeng)

Chapter 5 Protein Extraction

► Background Reading

Proteins were discovered by Jöns Jakob Berzelius (Fig. 5-1) in 1838 and are among the most actively-studied molecules in biochemistry. The word "protein" is derived from the Greek word "protas" meaning "of primary importance" suggesting the fundamental role of proteins in sustaining life. The word "protein" is defined as any of a group of complex high-molecular-weight organic compounds, essentially consisting of combinations of amino acids with peptide linkers containing carbon, hydrogen, oxygen, nitrogen, and usually, sulfur. Proteins are one class of bio-macromolecules (others include polysac, charides, lipids, and nucleic acids), making up the primary constituents of living things. Proteins have many important roles in the body and are part of every cell: muscles, connective tissue, blood-clotting factors, enzymes, antibodies, hormones, and bones. Proteins are common in foods such as poultry, fish, meat (e.g., beef, pork, lamb and veal), meat substitutes (soy), milk, eggs, cheese, nuts and legumes.

Fig. 5-1 Jöns Jakob Berzelius (1779-1848)

Principle of protein extraction

Protein extraction from tissues or cultured cells is the first step for many biochemical and analytical techniques (PAGE, western blotting, mass spectrometry, etc.) or protein purification. Efficient disruption and homogenization of animal tissues and cultured cells is required to ensure high protein yields. Several mechanical and chemical disruption techniques are available (Table 5-1). An efficient protocol for tissue and cell disruption is required to release proteins in a soluble form from their intracellular compartments. Disruption protocols should be as gentle as possible to the proteins because the extraction step is the starting point for all subseq-uent procedures. The success of tissue and cell disruption depends on a number of variables, such as the choice of buffers, the presence of protease inhibitors, and the osmolarity of the resuspension buffer. The conditions and constituents of the extraction buffer depend on the cell type, target protein, and its intended application.

Table 5-1 Various techniques for cell lysis

Technique	Priciple	Time of lysis	Example
Enzyme digestion	Digestion of cell wall leading to osmotic disruption of cell membrane	15-30 min	Gram positive bacteria
Osmotic shock lysis	Osmotic disruption of cell membrane	<5 min	Red blood cells
Hand homogenization	Cells are forced through narrow gap leading to disruption of cell membrane	10-15 min	Liver tissue
Blade homogenizer	Large cells are broken by chopping action	5-10 min	Animal tissue Plant tissue
Grinding with alumina or sand	Cell walls are ripped off by microroughness	5-15 min	Bacteria
Grinding with glass beads	Cell walls are ripped off by rapid vibration of glass	10-20 min	Bacteria
French press	Cells are forced through small orifice at very high pressure. Shear forces disrupt cells	10-30 min	Bacteria Plant cells
Sonication	Cell disruption by shear forces and cavitation caused by high-pressure sound wave	5-10 min	Bacteria

Proteins are extremely heterogeneous biological macromolecules. Their properties can be severely affected by small changes in hydrogen ion concentration, and thus a proper buffer with a stable pH is very important for maintaining the properties and biological activities of target proteins during protein extraction. Several factors are considered when choosing a buffer such as temperature sensitivity, interactions with other components (such as enzymes or metal ions), compatibility with different purification techniques, ultra-violet (UV) absorption, biological membrane permeability, and cost.

Proteolysis can be a major problem after extraction and at any stage of the protein purification process. This is because proteolysis may generate degraded proteins, partially retaining their biological activity or completely inactivate the desired proteins. This can lead to erroneous conclusions about the nature of the protein (such as size and structure). Several classes of proteases are present in cells, and fortunately, various protease inhibitors acting on each class of proteases are commercially available. Some commonly used protease inhibitors and their recommended concentrations are listed in Table 5-2. It is advisable to use a mixture of protease inhibitors when working with a new protein extract. Recently, a mixture of appropriately concentrated inhibitors became commercially available. For example, protease inhibitor cocktail, which is a mixture of several different protease inhibitors with different target protease specificities, is widely used to prevent protein degradation after lysing cells or tissues.

Detergents are generally added to buffers for extraction and purification of membrane proteins, which are usually insoluble in aqueous buffers. Detergents are a class of compounds characterized by their amphiphilic structure (both hydrophobic and hydrophilic). The three types of detergents include ionic, non-ionic, and zwitterionic. Ionic detergents denature proteins and, as such, are not used during protein extraction. Sodium dodecyl sulfate (SDS) and cetyltrimethylammonium bromide (CTAB) are examples of anionic and cationic detergents, respectively. Non-ionic detergents are generally used to isolate functional proteins, because they are far less denaturing than ionic detergents. Common examples of non-ionic detergents include Triton X-100 and Tween 20. Zwitter-ionic detergents are more effective than non-ionic detergents, as they prevent protein-protein interactions while causing less protein denaturation than ionic detergents. CHAPS is the most commonly used zwitterionic detergent. One of the most commonly used lysis buffers for extracting protein from cells and tissues is Radio Immuno Precipitation Assay (RIPA) buffer. The RIPA buffer is commercially available enables the extraction of membrane, nuclear, and cytoplasmic proteins, and is compatible with many applications, such as reporter assays, western blotting and immunoprecipitation.

Table 5-2　Common protease inhibitors

Inhibitor	Protease/phosphatase Inhibited	Final concentration in lysis buffer	Stock (store at −20℃)
Aprotinin	Trypsin, Chymotrypsin, Plasmin, Lysosomal	2 μg/mL	Dilute in water, 10 mg/mL. Do not re-use thawed aliquots
Leupeptin	Aspartic proteases 1	5-10 μg/mL	Dilute in water. Do not reuse once defrosted
Pepstatin A	Serine, Cysteine proteases	1 μg/mL	Dilute in methanol, 1 mM
PMSF	Metalloproteases that require Mg^{2+} and Mn^{2+}	1 mM	Dilute in ethanol. You can reuse the same aliquot
EDTA	Metalloproteases that require Ca^{2+} Serine/Threonine	5 mM	Dilute in ddH_2O, 0.5 M. Adjust pH to 8.0
EGTA	Phosphatases	1 mM	Adjust pH to 8.0
Na Fluoride		5 mM	Dilute in water. Do not reuse once defrosted
Na Orthovanadate		1 mM	Dilute in water. Do not reuse once defrosted

After obtaining proteins from cells and tissues, the total protein concentration should be determined. Total protein concentration is typically measured in the supernatant following extraction and clarification by centrifugation. Various methods can estimate the protein concentration, such as the bicinchoninic acid assay (BCA assay), Lowry protein assay, and Bradford protein assay. The BCA assay, also known as the Smith assay, is a biochemical assay for determining the total protein concentration in a solution (0.5 μg/mL to 1.5 mg/mL). The total protein concentration is measured by a color change of the sample solution from green to purple in proportion to protein concentration, which is then measured with colorimetric techniques.

In the present experiment, protein concentration will be determined by a spectrophotometer which directly measures the OD280 of protein samples. Benzene rings in tyrosine, phenylalanine and tryptophan residues contain conjugated double bonds, enabling proteins to absorb ultraviolet light. The absorption peak is at 280 nm, and optical density is directly proportional to protein concentration. In addition, the absorbance of a protein solution at 238 nm is directly proportional to the content of peptide bonds. These can be used to measure the protein content in a solution. Tissue from CMV-EGFP integrated transgenic pigs will be used in this experiment. CMV-EGFP integrated transgenic pigs were produced by somatic cell nuclear transferring (Fig. 5-2), and the EGFP reporter is highly expressed in heart and muscle.

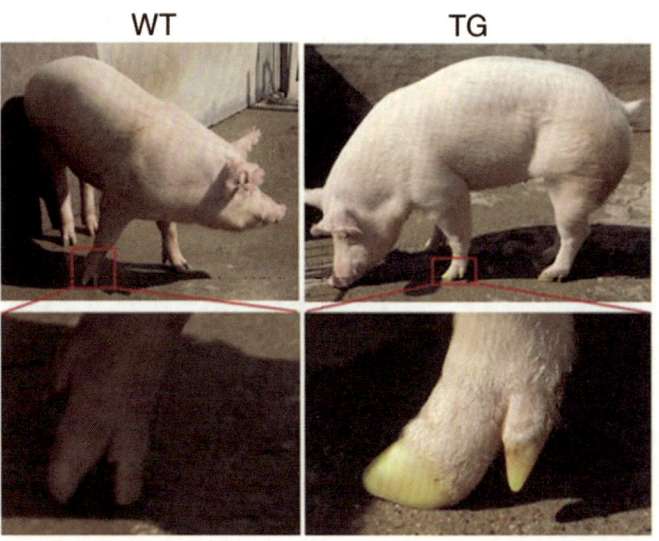

Fig. 5-2 EGFP expression of the transgenic pig T179 in the sunshine

WT: wild type Yorkshire; TG: transgenic Yorkshire T179

Section 1　Preparation of Protein

Objectives

- Learning the principle of protein extraction
- Isolation of protein from animal tissues
- Determination of protein concentration

Materials

Tissue: Muscle of EGFP-transgenic pig

Reagents: Ice-cold PBS; RIPA Buffer; PMSF; Liquid N_2

Suppliers: Microcentrifuge tubes; Mortar and pestle; Plastic cell scraper; Microcentrifuge tube

Equipment: Microcentrifuge; Spectrophotometer System; Ultrasonic Crushing Instrument

Activity protocol

(1) Add 50 μL fresh PMSF (100 mM) to 5 mL RIPA buffer, vortex and spin down.

(2) Chill mortar and pestle with liquid N_2.

(3) Remove the frozen pig muscle from 1.5 mL tube and transfer into chilled mortar.

(4) Grind muscle into a fine powder using the pestle.

(5) Transfer the powder into a chilled 5.0 mL tube and add 3-4 mL RIPA buffer with PMSF.

(6) Vortex the tissue suspensions vigorously for 15 s, and then put the samples on ice for 1 min. Repeat the vortex step for 2-3 times.

(7) Put the tube in a beaker with ice and further lyse the sample by sonication for 1 min (Pulse 02-04; Ampl 20%).

(8) Aliquot 0.5 mL tissue lysate in a 1.5 mL microcentrifuge tube.

(9) Centrifuge at 12,000 × g for 5 min in a microcentrifuge at 4 ℃.

(10) Gently remove the tubes from the centrifuge and place them on ice, then aspirate the supernatant and transfer it to a new tube kept on ice. Discard the pellet.

(11) Determine protein concentration by spectrophotometer.

Section 2　Know More

The Protein that Lights up Life Science

In the late 20th century, the discovery and application of a series of bioluminescent proteins represented by green fluorescent protein (GFP) provided a new tool for biology. Since then, fluorescent proteins have been used to trace living specimens, bringing a new experimental method to traditional biological research and making many biological fields leap into the quantitative research of dynamic living processes. Scientists have observed tiny structures and physiological processes of living cells more easily than before to understand complex phenomena such as embryo development and spread of cancer cells.

GFP is a protein composed of 238 amino acid residues (26.9 kDa) and was first isolated from the jellyfish Aequorea victoria. GFP exhibits bright green fluorescent when exposed to light in the blue to ultraviolet range. Enhanced GFP (EGFP) is a mutant of GFP with a 35-fold increase in fluorescence. This variant has mutations of Ser to Thr at amino acid 65 and Phe to Leu at position 64 and is encoded by a gene with optimized human codons. GFP has become well established as a gene expression marker and for protein targeting in intact cells and organisms. It has been used in modified forms to make biosensors and many animals have been created that express GFP as a proof-of-concept that a gene can be expressed throughout a given organism.

Information Box 5-1: GFP Amino Acid Sequence:

MSKGEELFTGVVPVLVELDGDVNG-
QKFSVSGEGEGDATYGKLTLNFICTTGK-
LPVPWPTLVTTFSYGVQCFSRYPDHMK-
QHDFFKSAMPEGYVQERTIFYKDDGNY-
KTRAEVKFEGDTLVNRIELKGIDFKEDG-
NILGHKMEYNYNSHNVYIMGDKPKNGI-
KVNFKIRHNIKDGSVQLADHYQQNTPIG-
DGPVLLPDNHYLSTQSALSKDPNEKRD-
HMILLEFVTAARITHGMDELYK

Because the GFP family of proteins dramatically changed the course of life sciences, the Royal Swedish Academy of Sciences awarded the 2008 Nobel Prize in chemistry to three scientists who made outstanding contributions to the discovery, expression, and development of GFP: Osamu Shimomura, Martin Chalfie and Roger Yonchien Tsien.

Osamu Shimomura was the first to isolate GFP from Aequorea victoria (Fig. 5-3), and found that the protein glows bright green under ultraviolet light[1].

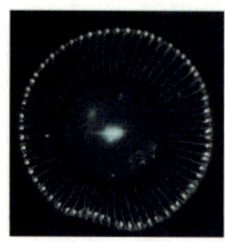
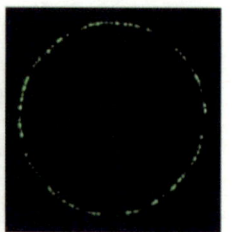

Fig. 5-3 Osamu Shimomura (1928-) and bioluminescent jellyfish

Martin Chalfie demonstrated the value of GFP as a luminous genetic marker for a variety of biological phenomena (Fig. 5-4). In an initial experiment, he used GFP to color six individual cells of C. elegans[2].

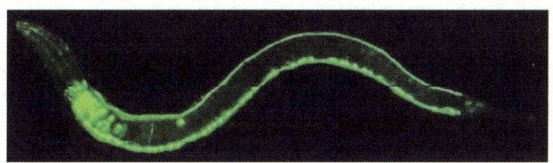

Fig. 5-4   Left: Martin Chalfie (1947-).   Right: C. elegans expressing GFP.

Tsien's main contribution was to understand the mechanism by which GFP fluoresces (Fig. 5-5). He developed mutants that start fluorescing faster, and are brighter, than wild type GFP (EGFP)[3], and fluoresce in other colors that can be used for labeling besides green, allowing scientists to apply different colors to various proteins and cells.

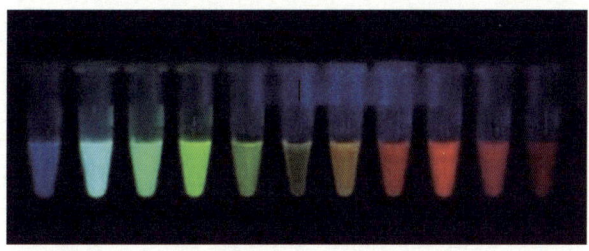

Fig. 5-5   Left: Roger Yonchien Tsien (1952-2016)
Right: Rainbow palette of GFP derived fluorescent proteins

GFP-based Techniques

The discovery of GFP has greatly promoted the development of modern biology and has a broad application prospect. The development of optical-molecular imaging based on novel functional fluorescent proteins (such as light-activated fluorescent proteins, light-converting fluorescent proteins, and REDOX sensitive GFP) provides more options for studying gene expression and protein function in living cells and even in living animals.

1. Applications of GFP in molecular biology

Applications in biology. As a marker molecule, GFP has been widely used in gene markers, protein markers, environmental microbial studies, parasitological studies, and the exploration of gene expression patterns in developmental biology studies, becoming a tool for the study of cellular molecules in living systems.

Applications in ecology. Introducing the GFP gene into target organisms or cloning the GFP gene into a virus, bacteria, or plasmid, could mark the target organism for studying ecology. For example, GFP was used to mark genetically engineered microorganisms to detect their survival and whereabouts in an aquatic environment, and to track the microbes.

Applications in medicine. The differentiation ability of adipose stem cells were confirmed using a GFP tracer. Enhanced GFP can also be used to screen suitable biomaterials. GFP is also used to study cytokine action and through *in vivo* dynamic observations. GFP is also widely used in gene transfer systems of the retina to regulate target genes and protect against photoreceptor apoptosis. In terms of immunological therapy, GFP is used as a tool for epidermal growth factor receptor (EGFR) research to track the subcellular localization of the second messenger protein, which can reveal its role in living cells and enable observation of the effect of signaling inhibitors, such as antibodies and small molecule tyrosine kinase inhibitors.

2. GFP as a reporter gene

GFP can be used as a reporter gene to detect transfection efficiency. After connecting the GFP gene to the promoter of a target gene, the expression level of that gene can be detected as the fluorescence intensity of GFP. Currently, this method has been widely used in both agrobacterium-mediated and gene-gun mediated plant genetic transformation and in living cells, transgenic embryos and animals, especially in spatiotemporal imaging of gene expression in living cells.

3. The GFP as a fusion tag

One of the most successful applications of GFP is its fusion as a tag onto the host protein to detect localization, migration, conformational changes, and intermolecular interactions of proteins, or to target certain organelles. In most cases, the GFP gene can combine with the heterologous gene at the N- or C-terminal to form chimeric genes encoding the fusion protein by standard molecular biological techniques, and the expression products maintain the biological activity of the endogenous protein, while showing similar fluorescence characteristics to the natural GFP. This characteristic of GFP can be used to detect protein localization and migration and to study the interaction and the conformational changes of proteins by relying on the fluorescence resonance energy transfer for detection.

4. Others

Wild-type GFP and many of its mutants can be used as biosensors to detect pH, and electrical potential, REDOX levels, and as a Ca^{2+} indicator in signal transduction in living cells. GFP is widely used in cell and drug screening, and is improved by fixed-point mutation, DNA-shuffling and other techniques. Various mutants with changes in fluorescence spectrum, quantum yield, solubility, and temperature sensitivity have been obtained.

References

[1] SHIMOMURA O, JOHNSON F H, SAIGA Y. Extraction, purification and properties of aequorin, a bioluminescent protein from the luminous hydromedusan, Aequorea [J]. J. Cell. Comp. Physiol., 1962, 59: 223-239.

[2] CHALFIE M, TU Y, EUSKIRCHEN G, WARD W W, et al. Green fluorescent protein as a marker for gene expression [J]. Science, 1994, 263: 802-805.

[3] HEIM R, CUBITT A, TSIEN R Y. Improved green fluorescence [J]. Nature, 1995, 373: 663-664.

Postlab Focus Questions

1. What is the function of PMSF in RIPA buffer? Why should it be freshly added into RIPA buffer just before protein extraction?

2. Why should the sample be put on ice during sonication?

3. Why was RIPA buffer with PMSF used as the "blank" when measuring protein concentrations?

(Written by Sun Caiyun)

Chapter 6 Protein Analysis

Background Reading

The separation of macromolecules in an electric field is called electrophoresis. Proteins are commonly separated by an electrophoresis technique that uses a discontinuous polyacrylamide gel as a support medium and sodium dodecyl sulfate (SDS) to denature the proteins. This method is called sodium dodecyl sulfate polyacrylamide gel electrophoresis (SDS-PAGE). The most commonly used system is also called the Laemmli method after U. K. Laemmli, who was the first to publish a scientific study employing SDS-PAGE in a scientific study.

Proteins, unlike nucleic acids, can have varying charges and complex shapes; therefore, they may not migrate on the gel at similar rates, or at all, when a negative to positive electromotive force (EMF) is placed on the sample. Proteins, therefore, are typically denatured in the presence of a detergent, such as sodium dodecyl sulfate/sodium dodecyl phosphate (SDS/SDP), that coats the proteins with a negative charge. SDS is an anionic detergent, meaning that when dissolved its molecules have a net negative charge within a wide pH range. A polypeptide chain binds SDS in proportion to its relative molecular mass. The negative charges on SDS destroy most of the complex protein structures, and are strongly attracted toward an anode (positively-charged electrode) in an electric field. SDS-PAGE maintains polypeptides in a denatured state after treatment with strong reducing agents to remove secondary and tertiary structure (e. g. disulfide bonds [S-S] to sulfhydryl groups [SH and SH]) and thus allows proteins to be separated by their molecular weight. Sampled proteins become covered in the negatively charged SDS and move to the positively charged electrode through the acrylamide mesh of the gel. Smaller proteins migrate faster through this mesh, thus separating the proteins according to size (usually measured in kilodaltons, kDa). The concentration of acrylamide determines the resolution of the gel; the greater the acrylamide concentration, the better the resolution of lowering molecular weight proteins. Alternatively, lowering the acrylamide concentrations provides better resolution of higher molecular weight proteins. Proteins travel only in one dimension along the gel for most blots.

Protein separation by SDS-PAGE can be used to estimate relative molecular mass, to determine the relative abundance of major proteins in a sample, and to determine the distribution of proteins among fractions. The purity of protein samples can be assessed and the progress of a fractionation or purification procedure can be followed. Different staining methods can be used to detect rare proteins and to learn something about their biochemical properties. Specialized techniques such as Western blotting, two-dimensional electrophoresis, and peptide mapping can be used to detect extremely scarce gene products, to find similarities among them, and to detect and separate isoenzymes of proteins.

The Western blotting is an analytical technique used to detect specific proteins in a given sample of tissue homogenate or cell extract. The method originated from the laboratory of George Stark at Stanford. The name "Western blot" was given to the technique by W. Neal Burnette and is a play on the name Southern blot, a technique for DNA detection previously developed by Edwin Southern. Western blotting uses gel electrophoresis to separate native or denatured proteins by the length of the polypeptide (denaturing conditions) or by the 3-D structure of the protein (native/non-denaturing conditions). The proteins are then transferred to a membrane (typically nitrocellulose or polyvinylidene fluoride), where they are probed (detected) using antibodies specific to the target protein. This method is used in the fields of molecular biology, biochemistry, immunogenetics and other molecular biology disciplines.

Gel electrophoresis

Soluble antigens (the target protein) may be separated by SDS-PAGE based on their molecular weight, size and charge (non-denaturing gel electrophoresis or isoelectric point). SDS-PAGE is the most widely used technique to separate proteins in various samples of mixture (Fig. 6-1). In a polyacrylamide gel, proteins migrate to the anode (negatively charged) under denaturing conditions. In SDS-PAGE, the SDS detergent and a heating step ensure that the electrophoretic mobility is only determined by the molecular weight of proteins. A complete protein electrophoresis system contains: a tank, lid with power cables, electrode assembly, cell buffer dam, casting stands, casting frames, combs (usually 10-well or 15-well), and glass plates (thickness 0.75 mm or 1.0 mm or 1.5 mm).

Protein gels in a single electrophoresis run are divided into a stacking gel and a separating gel. The stacking gel (5% acrylamide) is poured on top of the separating gel, and a gel comb is inserted in the stacking gel to create the wells. The acrylamide percentage in the separating gel depends on the size of the target protein in the sample (Table 6-1).

Table 6-1　Acrylamide percentage in the separating gel

Acrylamide %	M. W. Range
7%	50-500 kDa
10%	20-300 kDa
12%	10-200 kDa
15%	3-100 kDa

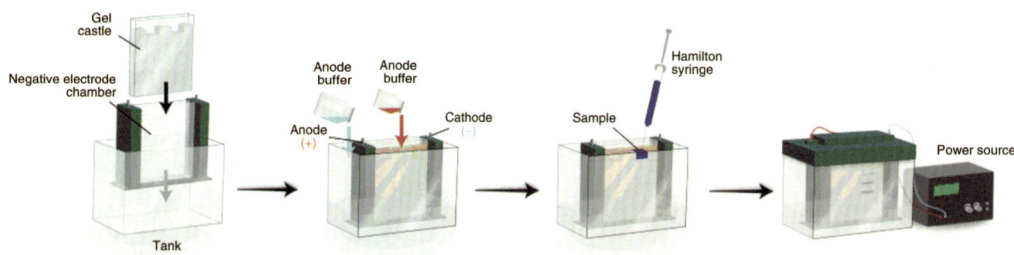

Fig. 6-1　Scheme of SDS-PAGE gel electrophoresis

Transferring

To make the proteins accessible to antibody detection, after the SDS-PAGE, the proteins are transferred from the gel to a nitrocellulose or polyvinylidene fluoride (PVDF) membrane. The arrangement of membrane and gel is shown as follows (Fig. 6-2): filter paper on top, followed by the gel, the Western blot membrane, and finally another filter paper on the bottom. The electric current pulls proteins from the gel into the PVDF or nitrocellulose membrane. The proteins move from within the gel onto the membrane while maintaining the organization they had within the gel. As a result

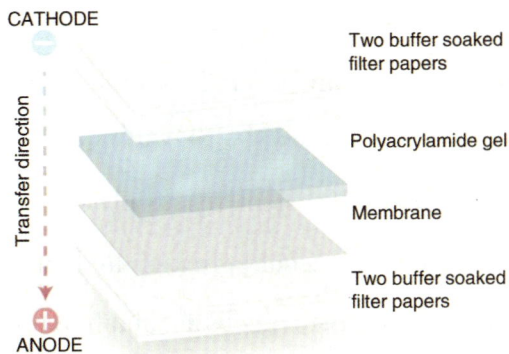

Fig. 6-2 Arrangement of membrane and gel for protein transferring

of this "blotting" process, the proteins are exposed on a thin surface layer of the membrane for detection. Both varieties of membrane are chosen for their non-specific protein binding properties (i.e. all proteins are bound equally well). Protein binding is based on hydrophobic interactions, and charged interactions between the membrane and protein. Nitrocellulose membranes are cheaper than PVDF, but are far more fragile and do not stand up well to repeat probing. The uniformity and overall effectiveness of protein transfer from the gel to the membrane can be checked by staining the gel with Coomassie or Ponceau S dyes.

Blocking

Given that the membrane has been chosen for its ability to bind protein, and both antibodies and targets are proteins, steps must be taken to prevent interactions between the membrane and the antibody used for detecting the target protein. Non-specific binding is blocked by placing the membrane in a dilute solution of protein, typically bovine serum albumin (BSA) or non-fat dry milk, with a minute percentage of detergent, such as Tween 20. The protein in the dilute solution attaches to the membrane where the target proteins have not. Thus, when the antibody is added, there is no room on the membrane for it to attach other than on the specific binding sites of the target protein. This reduces "noise" in the final product of the Western blot, leading to clearer results and eliminating false positives.

Detection

During the detection process, the membrane is "probed" for the protein of interest with a modified antibody linked to a reporter enzyme that drives a colorimetric reaction to produce a color when exposed to an appropriate substrate. For a variety of reasons, this traditionally takes place in a two-step process, although there are now one-step detection methods available for certain applications.

Primary antibody

Antibodies are generated when a host species or immune cell culture is exposed to the protein of interest. Normally, this is part of the immune response, whereas here they are harvested and used as sensitive and specific detection tools that directly bind the protein of interest.

After blocking, a dilute solution of primary antibody (generally between 0.5 and 5 μg/mL) is incubated with the membrane under gentle agitation. Typically, the solution is composed of buffered saline solution with a small percentage of detergent, and sometimes with powdered milk or BSA. The antibody solution and the membrane can be sealed and incubated together for anywhere from 30 min to overnight. It can also be incubated at different temperatures, with warmer temperatures being associated with more binding, both specific (to the target protein, the "signal") and non-specific ("noise").

Secondary antibody

After rinsing the membrane to remove unbound primary antibody, the membrane is exposed to another antibody, directed at a species-specific portion of the primary antibody. This is known as a secondary antibody, and tends to be referred to as "anti-mouse," "anti-goat," etc. due to its targeting properties. Antibodies originate from animal sources (or animal sourced hybridoma cultures); an anti-mouse secondary will bind to essentially any mouse-sourced primary antibody. This allows some cost savings by allowing an entire lab to share a single source of mass-produced secondary antibody and provides far more consistent results. The secondary antibody is usually linked to biotin or a reporter enzyme, such as alkaline phosphatase or horseradish peroxidase. Several secondary antibodies will bind to one primary antibody to enhance the signal. Most commonly, a horseradish peroxidase-linked secondary is used in conjunction with a chemiluminescent agent, and the reaction product produces luminescence proportional to the amount of protein.

Analysis

After unbound probes are washed away, the Western blot is ready for detection of probes that are labeled and bound to the protein of interest. In practical terms, not all Western blots reveal protein at only one band in a membrane. Size approximations are made by comparing the stained bands to that of the marker or ladder loaded during electrophoresis. The process is repeated for a structural protein, such as actin or tubulin that should not change between samples. The amount of target protein is normalized to the structural protein to control for variability between groups. This practice corrects for the total amount of protein on the membrane in case of errors or incomplete transfers.

Colorimetric detection

The colorimetric detection depends on incubation of the Western blot with a substrate that reacts with the reporter enzyme (such as peroxidase) bound to the secondary antibody. This converts the soluble dye into an insoluble form of a different color that precipitates next to the enzyme, thereby staining the membrane. Development of the blot is stopped by washing away the soluble dye. Protein levels are evaluated through densitometry (i. e. intensity of the stain) or spectrophotometry.

3, 3′-Diaminobenzidine (DAB) is a widely used chromogen for immunohistochemical staining and immunoblotting. In the presence of peroxidase enzyme, DAB produces a brown precipitate that is insoluble in alcohol and xylene. This product comes in a two-component system consisting of a liquid stable DAB chromogen and DAB substrate buffer. The color of the chromogen solution can vary from colorless to pale brown.

Chemiluminescent detection

Chemiluminescent detection depends on incubation of the Western blot with a substrate that will luminesce when exposed to the reporter on the secondary antibody. The light is then detected by photographic film, and more recently by CCD cameras which capture a digital image of the Western blot. The image is analyzed by densitometry to evaluate the relative amount of protein staining and quantify the results in terms of optical density.

Since 1988, enhanced chemiluminescence (ECL) has become one of the most common detection methods in Western blotting. In this method, the secondary antibody is conjugated to the enzyme horseradish peroxidase. Once bound to the membrane, the secondary antibody is detected by incubating the blot with a solution containing an HRP substrate that generates a light-emitting product after reaction with HRP. The chemiluminescent signal can be detected by exposing the blot to X-ray film or by imaging with a CCD camera.

Radioactive detection

Radioactive labels do not require enzyme substrates, but rather enable medical X-ray film to be placed directly against the Western blot to develop as it is exposed to the label, creating dark regions corresponding to the protein bands of interest. The importance of radioactive detection is on the decline due to its hazardous radiation, cost, high health and safety risks, and the development of ECL as a useful alternative.

Secondary probing

One major difference between nitrocellulose and PVDF membranes relates to the ability of each to support "stripping" antibodies off and reusing the membrane for subsequent antibody probes. While there are well-established protocols available for stripping nitrocellulose membranes, the sturdier PVDF membranes allow for easier stripping and more reuse before background noise limits experiments. Another difference is that unlike nitrocellulose, PVDF must be soaked in 99% methanol before use. PVDF membranes also tend to be thicker and more resistant to damage during use.

Section 1 SDS-PAGE and Western Blotting

Objectives

❖ Understanding the theory of SDS-PAGE and Western blotting

❖ Learning basic manipulation and applications

Materials

Reagents: 30% Acrylamide; 10% SDS; 10% APS (make fresh each time); TEMED; 1.5 M Tris-HCl, pH 8.8 (separating gel); 1.0 M Tris-HCl, pH 6.8 (stacking gel); 5 × SDS running buffer; SDS sample loading buffer; Transfer Buffer; 1 × TBS/0.1% Tween 20; Blotting buffer; Primary and secondary antibodies; ECL and DAB Western blotting detection reagents.

Suppliers: Pippet tips; PVDF membrane; Whatman 3MM paper; Sponge pad; Ice boxes; Forceps; Scissor; Small plastic or glass container; Shallow tray.

Equipment: apparatus of SDS-PAGE (Fig. 6-3); Electroblotting Apparatus (Fig. 6-4); Power suppler; Shaker; Heater; Refrigerator; Automatic chemilumi-nescence image analysis system.

Fig. 6-3 Apparatus for SDS-PAGE

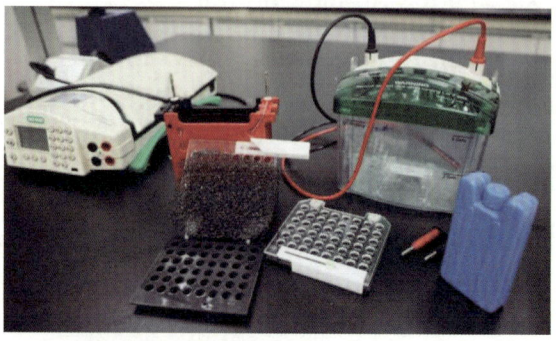

Fig. 6-4 Apparatus for protein transferring

Activity protocol

Part 1. SDS-PAGE

(1) Clamp two glass plates (1.5 mm thickness) in the casting frames on the casting stands (Fig. 6-5).

(2) Prepare 20 mL of 12% separating gel solution (Table 6-2) and pour it (7.5 mL for each gel) into the gap between the glass plates (Fig. 6-6). Then pour water or isopropanol into the gap until it overflows. Wait 30-40 min to let it condense.

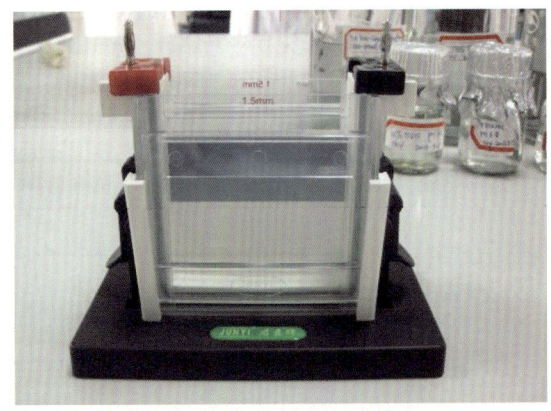

Fig. 6-5 Setting up the casting frames

Table 6-2 Formula for separating gel (10 mL)

Acylamide percentage	6%	8%	10%	12%	15%
H$_2$O	5.2 mL	4.6 mL	3.8 mL	3.2 mL	2.2 mL
30% (w/v) Acrylamide	2 mL	2.6 mL	3.4 mL	4 mL	5 mL
1.5 M Tris-HCl (pH = 8.8)	2.6 mL	2.6 mL	2.6 mL	2.6 mL	2.6 mL
10% (w/v) SDS	100 μL	100 μL	100 μL	100 μL	100 μL
10% (w/v) Ammonium persulfate (AP)	100 μL	100 μL	100 μL	100 μL	100 μL
TEMED	10 μL	10 μL	10 μL	10 μL	10 μL

Note: *For each group, two gels should be prepared. One to detect the target protein (GFP), and the other for the internal control (GAPDH).*

(3) Prepare 10 mL of 5% stacking gel solution (Table 6-3). Pour out the water and you can see the separating gel remaining. Then pour the stacking gel (2.5 mL for each gel) until it overflows. Then insert the comb and wait 20-30 min to let it condense.

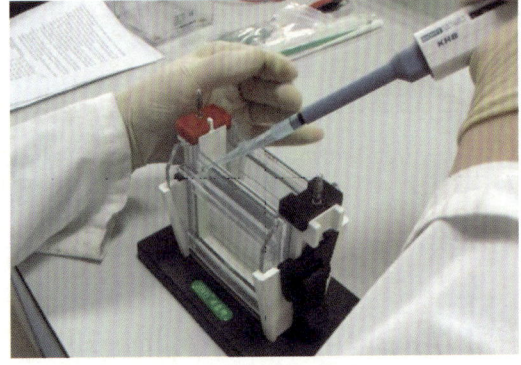

Fig. 6-6 Preparing the separating and stacking gel

Table 6-3 Formula for 5% stacking gel (10 mL)

H_2O	6.8 mL
30% (w/v) Acrylamide	1.7 mL
1.0 M Tris-HCl, pH 6.8	1.25 mL
10% (w/v) SDS	100 μL
10% (w/v) Ammonium persulfate (AP)	100 μL
TEMED	10 μL

(4) Ensure the stacking gel has completely condensed, then take out the comb. Take the glass plates out of the casting frame and set them in the cell buffer dam. Pour the running buffer into the inner chamber until the buffer surface reaches the required level in the outer chamber (Fig. 6-7).

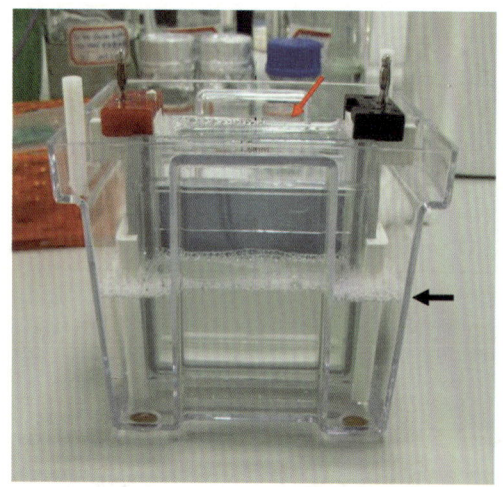

Fig. 6-7 Buffer surface level in inner chamber (red arrow) and outer chamber (black arrow)

(5) Prepare the samples: mix 30 μL of samples with 10 μL 4 × loading buffer together, vortex gently and spin down. Heat the mixture in boiling water for 10 min, then put on ice for at least 5 min.

(6) Load 10 μL protein marker into the first lane, then load 10-30 μg sample into the wells. Connect the electrodes and run the electrophoresis for 60-90 min (80 V for stacking gel and 120 V for separating gel).

(7) Typically run about 1.5 h at 120 V for a 12% separating gel. For a higher percentage of acrylamide, the time taken would be longer (Fig. 6-8).

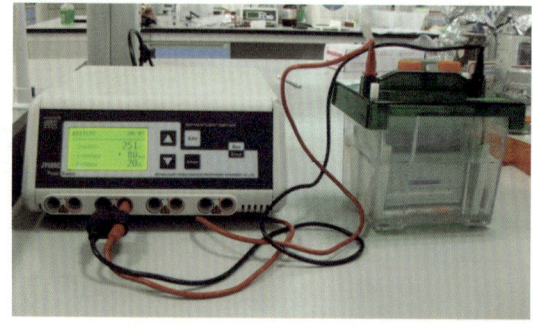

Fig. 6-8 SDS-PAGE gel electrophoresis

Part 2. Transfer of protein to PVDF membrane

(1) Prepare fresh 1 × transfer buffer.

(2) Cut off the stacking gel and nick the top left-hand corner of the resolving gel for orientation.

(3) Measure the dimensions of the gel and note the positions of the ladder bands.

(4) Transfer the gel, while still attached to the glass plate, to a box containing transfer buffer and peel the gel off the plate gently with a spatula (Fig. 6-9).

(5) Agitate 15-20 min at room temperature to remove salts and SDS.

(6) Cut a piece of PVDF membrane to the size of the gel and mark and/or clip one corner as the top left-hand corner. Handle only with flat forceps.

(7) Activate the membrane in methanol for 1 min and then immerse it in transfer buffer for 10-15 min.

(8) Cut four pieces of 3 mm filter paper to the dimensions of the gel.

(9) Open a gel holder cassette, red side down.

(10) Soak a fiber pad with transfer buffer and place in the center of the red side.

(11) Soak two pieces of filter paper with transfer buffer and place on top of the fiber pad.

(12) Place the PVDF membrane on top of filter paper.

(13) Roll out any bubbles with a glass tube and add 3 mL of transfer buffer to the top.

(14) Pour out old transfer buffer and add fresh transfer buffer to the gel.

(15) Take the gel out with a glass plate.

(16) Place the gel on top of the PVDF membrane.

(17) Roll out any bubbles with a glass tube and add 3 mL of transfer buffer to the top.

(18) Soak two pieces of filter paper with transfer buffer and place on top of the gel.

(19) Roll out any bubbles with a glass tube and add 3 mL of transfer buffer to the top.

(20) Close the gel holder cassette.

(21) Place in a transfer tank (orient the red and black sides of the cassette with the red and black panels of the electrode) and fill with transfer buffer (−800 mL).

(22) Place the tank in a styrofoam box with ice.

(23) Run at 60 V for 30-40 min (Fig. 6-10).

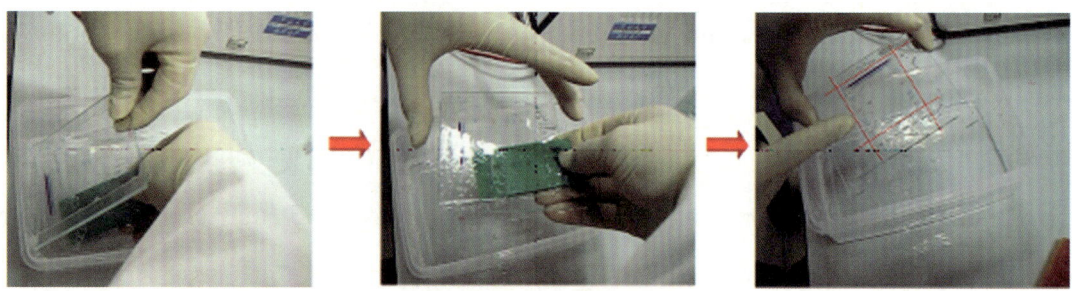

Fig. 6-9 Preparation of gel for transfer

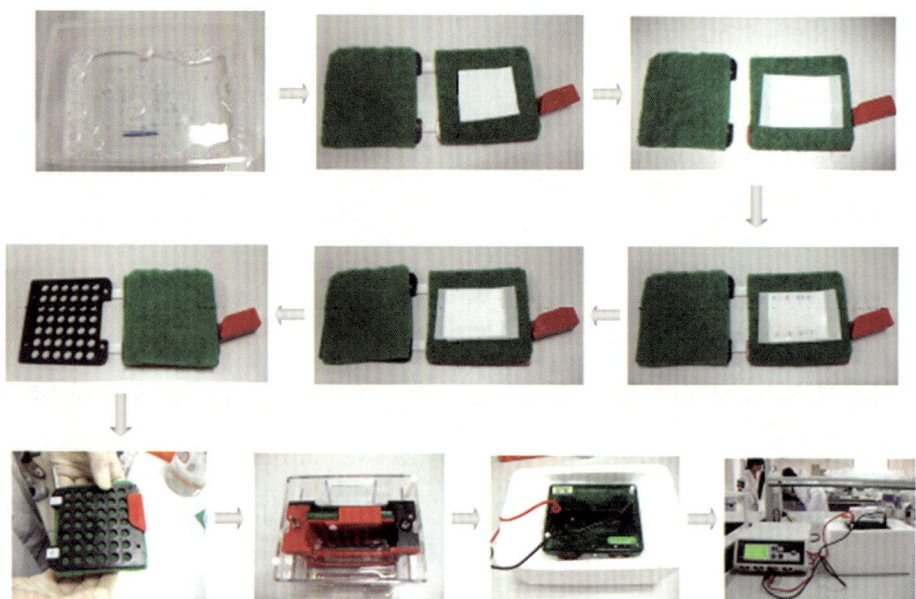

Fig. 6-10 Transferring protein to PVDF membrane

Part 3. Immunodetection

(1) Prepare 10 mL blocking solution [5% (W/V) non-fat dry milk, 0.5 g milk in 10 mL 1×TBST] for a piece of membrane.

(2) After completing the transfer, place the membrane in a box with 10 mL 1×TBST, gently shake the box several times, and then discard the TBST buffer.

(3) Add 10 mL blocking solution to the box and incubate the membrane in blocking solution at room temperature for 1 h with agitation.

(4) Discard blocking solution and wash the membrane with 10 mL 1×TBST for 10 min with agitation. Repeat this step two more times.

(5) Prepare the primary antibodies (1° ab) for EGFP and GAPDH.

◇1 μL GFP primary antibody is added to 1 mL 1×TBST with 5% (W/V) non-fat dry milk (1:1000 dilution).

◇1 μL GAPDH primary antibody is added to 1 mL 1 × TBST with 5% (W/V) non-fat dry milk (1 : 1000 dilution).

(6) Pour 15 mL water into a box (for holding 10 μL or 20 μL tips), then cover the interface layer (with holes). Put a piece of rectangular parafilm on the layer. Place membranes on the parafilm and add 1 mL 1° ab to the membrane drop by drop. Then gently put a small piece of parafilm on the membrane. Cover the box and incubate in a refrigerator at 4 ℃ overnight (Fig. 6-11).

(7) Wash membranes with 10 mL 1 × TBST for 10 min with agitation and repeat two times.

(8) During the washing step, prepare the secondary antibodies (2° ab) for GFP and GAPDH.

◇5 μL HRP-conjugated anti-Mouse IgG antibody is added to 10 mL 1 × TBST with 5% (W/V) non-fat dry milk (1 : 2000) for the membrane incubated with the 1° ab for GFP.

◇5 μL HRP-conjugated anti-Rabbit IgG antibody is added to 10 mL 1 × TBST with 5% (W/V) non-fat dry milk (1 : 2000) for the membrane incubated with the 1° ab for GAPDH.

(9) Discard the wash buffer (1 × TBST) and incubate the membranes with 10 mL of their corresponding 2° ab in TBST with agitation at RT for 1 h.

(10) Wash as in step 7.

(11) Rinse twice with ddH_2O for 1-2 min to remove the Tween-20.

(12) Place membranes on plastic wrap.

For ECL detection

(13) Open Automatic Chemiluminescence Image Analysis System in advance to make sure that the temperature of this system's CCD decreases at least −30 ℃.

(14) Mix ECL detection reagents thoroughly according to the introduction. Cover the tube with aluminum foil after mixing.

(15) Put the membrane in the automatic chemiluminescence image analysis system detection platform. Add 0.5 mL mixed substrate to the membrane and incubate for 1 min.

(16) Select optimal exposure time and take photos. In the present experiment, try 1 min as the initial exposure time.

For DAB detection

(13) Add DAB reagents according to the introduction. Cover the tube with the mixed substrate with aluminum foil.

(14) Add the DAB substrate solution to the membrane and incubate until the desired development is achieved. Typical incubations for the present experiment range from 2-5 min.

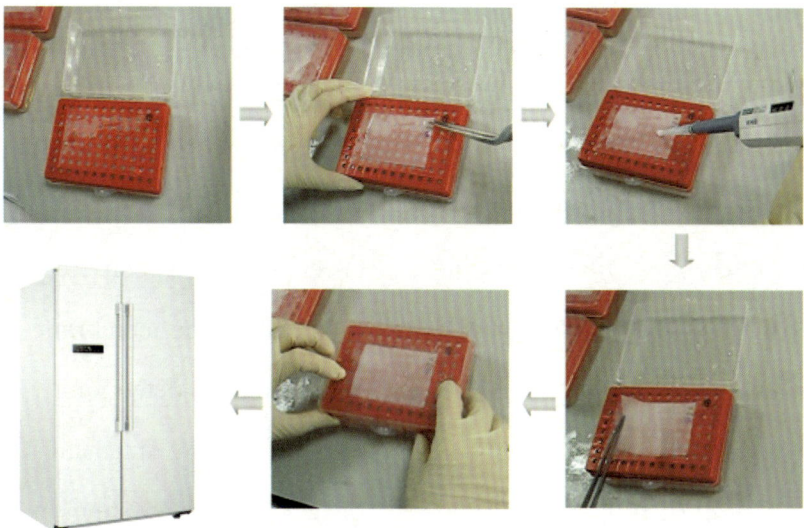

Fig. 6-11 Procedure for primary antibody incubation

Section 2 Know More

Proteomics

Proteomics is the large-scale study of proteomes. The "proteome" can be defined as the overall protein content of a cell and is characterized with regard to their localization, interactions, post-translational modifications, and turnover at a particular time[1]. The first proteomic studies began in 1975, after the introduction of the 2-D gel and mapping of the proteins from the bacterium *E. coli*. The term "proteomics" was firstly used by Marc Wilkins in 1996 to denote the "PROTein complement of a genOME". After genomics and transcriptomics, proteomics is the next step in the study of biological systems. It is more complicated than genomics because an organism's genome is more or less constant whereas proteomes differ from cell to cell and from time to time. Distinct genes are expressed in different cell types, meaning that even the basic set of proteins produced in a cell need to be identified. In the past, this phenomenon was assessed by RNA analysis, but this was shown to lack correlation with protein content. Now it is known that mRNA is not always translated into protein, and the amount of protein produced for a given amount of mRNA depends on the gene it is transcribed from and on the current physiological state of the cell. Proteomics confirms the presence of the protein as well as directly quantifies the amount present.

Many different areas of study are explored

through proteomics. Amongst them are protein-protein interactions, protein function, protein modifications, and protein localization. The fundamental goal of proteomics is not only to identify all of the proteins in a cell, but also to generate a complete three-dimensional map of the cell indicating each protein's exact location. There are three major types of proteomics, including structural proteomics, expression proteomics, and functional proteomics. The goal of structural proteomics is to map out the proteins present in a specific cellular organelle or the structure of protein complexes[2]. Structural analysis can aid in identifying the functions of newly discovered genes, show where drugs bind to proteins and where proteins interact with each other. Technologies employed in structural proteomics include X-ray crystallography and nuclear magnetic resonance spectroscopy. The quantitative study of protein expression between samples that differ by certain variables is known as expression proteomics. This type of proteomics can help identify the main proteins found in a particular sample, and proteins differentially expressed in related samples, e. g. when comparing diseased and healthy tissues. Technologies such as 2D-PAGE and mass spectrometry are used in expression proteomics. Functional proteomics represents a broad term encompassing many specific, directed proteomics methodologies. The characterization of protein-protein interactions is used to determine protein functions and to demonstrate how proteins assemble in larger complex-es. In some cases, specific sub-proteomes are isolated by affinity chromatography for additional analysis (Fig. 6-12).

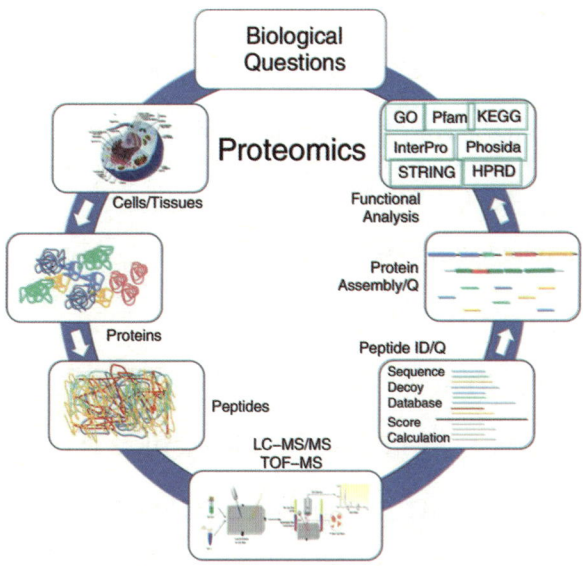

Fig. 6-12 The proteomic approach

Several high-throughput technologies have been developed to investigate proteomes in depth. The most commonly applied are mass spectrometry (MS)-based techniques, such as Tandem-MS, and gel-based techniques, such as differential in-gel electrophoresis (DIGE). These high-throughput technologies generate huge amounts of data. Databases are critical for recording and carefully storing this data, allowing researchers to make connections between their results and existing knowledge. Bioinformatics enable faster analysis and data storage. A good source to find lists of current programs and databases is on the ExPASy bioinformatics resource portal. The applications of bioinformatics-based proteomics include disease diagnosis, biomarker identification, and many more[3].

A major practical application of proteomics is the identification of potential new drugs for treating diseases. This relies on genome and proteome information to identify proteins associated with a disease, and then uses computer software to search for new drugs to target them. For example, if a certain protein is implicated in a disease, the 3D structure provides information on how to design drugs to interfere with that protein's function. A molecule that fits in the active site of an enzyme, but cannot be released by the enzyme, inactivates the enzyme. This is the basis of new drug-discovery tools, which aim to find new drugs to inactivate proteins involved in disease.

References

[1] ANDERSON N L, ANDERSON N G. Proteome and proteomics: new technologies, new concepts, and new words[J]. Electrophoresis, 1998, 19: 1853-1861.

[2] BLACKSTOCK W P, WEIR M P. Proteomics: quantitative and physical mapping of cellular proteins[J]. Trends biotechnol, 1999, 17: 121-127.

[3] Bilal A, Madiha B, et. al. Proteomics: technologies and their applications[J]. Journal of chromatographic science, 2017, 55: 182-196.

Postlab Focus Questions

1. How do the stacking gel and separating gel work together to separate proteins according to their molecular weight in SDS-PAGE?

2. What is the function of loading buffer in Western blot?

3. Please elucidate the function of the secondary antibody during Western blotting.

4. What are the main differences in detection methods between ECL and DAB?

(Written by Sun Caiyun)

Chapter 7　Animal Cell Culture

Background Reading

Cell culture is a technique in which cells are separated from the body, and then live and grow *in vitro* under appropriate conditions. Initially, the technique was designed to avoid the effects of homeostasis and the holistic effects of experimental stress. The main advantage is that it can be used in experiments with a relatively consistent set of cloned cells and reduces the number of laboratory animals used for studies. After more than a century of development, cell culture has been widely used in biology, medicine, drug development, and other fields, and has become one of the most important basic experimental technologies.

Jolly showed, for the first time, that cells can survive and divide *in vitro*. The practice of cell culture originated in 1907, when Ross Harrison (Fig. 7-1) isolated frog embryonic nerve fibers, maintained them using lymphatic fluid as the medium, and observed the process of nerve cell protuberant growth.

Fig. 7-1　Ross Harrison (1870-1959)

In the late 1940s, Enders, Weller, and Robbins grew poliomyelitis virus in culture, paving the way for testing many chemicals and antibiotics that affect viral multiplication in living host cells. The significance of animal cell culture was increased when viruses were used to produce vaccines in animal cell cultures in the late 1940s.

L929 is the earliest cloned cell line and was isolated from mouse L-cells in 1948 by Sanford et al. using capillary cloning. In 1952, Gey established the first continuous human cell line, the HeLa cell line. Cell culture techniques developed rapidly in the 1950s due to the development and use of trypsin, antibiotics, and various standard culture techniques.

Types of cell culture

Most cell lines begin as primary cultures originating from a piece of minced or enzyme-dispersed tissue. Primary cultures, as mixtures of several cell types, retain the characteristics of their source tissue (Fig. 7-2).

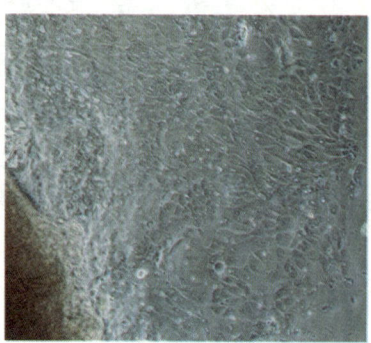

Fig. 7-2 Primary cell culture

After a period of time, primary cultures will reach confluency, i. e. when all available space of the culture vessel is covered due to cellular expansion. At this point, the culture needs to be disaggregated (usually with proteolytic enzymes like trypsin) into individual cells and subcultured (split, passaged, or transferred). With each subsequent subculture, the cellular population becomes more homogeneous as the faster growing cells predominate. Cells with desired properties can also be selected from the culture by cloning. Diploid cell lines rarely progress beyond a few population doublings because they have a finite replicative capacity and begin to slow down and eventually stop dividing after 20 to 50 population doublings. Recent evidence has suggested that some cellular senescence observed in cell culture may be due to inappropriate culture conditions as opposed to a predetermined replicative senescence. However, other data supports replicative senescence for cells of some species (notably human) even when grown in improved culture conditions. This senescence is mediated by the shortening of the ends of the chromosomes (telomeres) with each cell division.

In contrast, continuous (or immortalized) cell lines have infinite replicative capacity. These lines are derived through immortalization or transformation by any one of a number of means. Many continuous cell lines have been derived from tumor tissue. Cell lines either grow attached to a surface (anchorage dependent) or in suspension (anchorage independent). As cells grow and divide in a monolayer or in suspension, they usually follow a characteristic growth pattern composed of four phases: lag, log or exponential, stationary or plateau, and decline (Fig. 7-3).

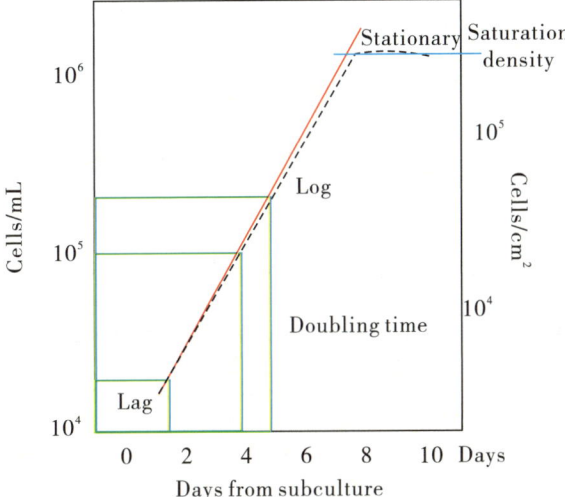

Fig. 7-3 Growth curve for cells grown in culture. Cells should be subcultured while still in the exponential phase

Lag phase — Immediately after seeding the culture vessel: the cells grow slowly while recovering from the stress of subculturing.

Log or exponential phase — The cells enter a period of exponential growth that lasts until the entire growth surface is occupied or the cell concentration exceeds the capacity of the medium.

Stationary phase — Cell proliferation slows and stops.

Decline phase — If the culture medium is not replaced and the cell number is not reduced, the cells lose viability and their number decreases.

To ensure viability, genetic stability, and phenotypic stability, cell lines need to be maintained in the exponential phase. This means that they need to be subcultured on a regular basis before entering the stationary growth phase, before a monolayer becomes 100% confluent, or before a suspension reaches its maximum recommended cell density. Generating a growth curve for each cell line is useful to determine the growth characteristics of the cell line.

Culture conditions

Temperature

Most animal cell lines require 37 ℃ for optimum growth. Insect and amphibian cells require lower temperatures (such as 28 ℃), as do some animal cell lines with temperature sensitive phenotypic characteristics. While cultured cells can withstand considerable drops in temperature, and most can survive for several days at 4 ℃, few can tolerate even a few hours at more than 2 ℃ above their optimal temperature.

Cell culture media

Cell culture media are complex mixtures of salts, carbohydrates, vitamins, amino acids, metabolic precursors, growth factors, hormones, and trace elements. The requirements for these components vary among cell lines, leading, in part, to the extensive number of medium formulations.

Carbohydrates are primarily supplied as glucose. In some instances, glucose is replaced with galactose to decrease lactic acid build-up, as galactose is metabolized at a slower rate. Other carbon sources include amino acids (particularly L-glutamine) and pyruvate. In addition to nutrients, the medium helps maintain the pH and osmolality in a culture system. The pH is maintained by one or more buffering systems; CO_2/sodium bicarbonate, phosphate, and HEPES are the most common. Serum will also buffer a complete medium. Phenol red, a pH indicator, is added to media to colorimetrically monitor changes in pH.

Media ingredients

Sodium bicarbonate and buffering

Cells produce and require small amounts of carbon dioxide (CO_2) for growth and survival. In culture media, dissolved CO_2 is in equilibrium with bicarbonate ions and many media formulations take advantage of this CO_2/bicarbonate reaction to buffer the pH of the medium. CO_2 dissolves freely into the medium and reacts with water to form carbonic acid. As the cells metabolize and produce more CO_2, the pH of the medium decreases. The optimal pH range of 7.2 to 7.4 can be maintained by supplementing the medium with sodium bicarbonate and regulating the atmospheric level of CO_2.

HEPES buffer

HEPES and other organic buffers can be used with many cell lines to effectively buffer the pH of the medium. Indeed, some standard medium formulations include HEPES. However, this compound can be toxic, especially for some differentiated cell types. HEPES has been shown to greatly increase the media's sensitivity to the phototoxic effects induced by exposure to fluorescent light.

L-Glutamine

L-Glutamine is an essential amino acid required by virtually all mammalian and insect cells grown in culture. It is used for protein production, as an energy source, and in nucleic acid metabolism. It is also more labile in liquid cell culture media than other amino acids.

The rate and extent of L-glutamine degradation are related to storage temperatures, age of the product, and pH. Because of the labile nature of L-glutamine, it is often omitted from commercial liquid media preparations to lengthen the product shelf life. In these cases, it must be aseptically added prior to use. L-Glutamine is not as labile in its dry form and most powdered media formulations include it. In some cases, the addition of L-glutamine to complete cell culture medium can extend the usable life of the medium. However, please use caution while adding more L-glutamine than is called for in the original medium formulation. L-glutamine degradation results in the build-up of ammonia which can have a deleterious effect on some cell lines. For most cell lines, ammonia toxicity is more critical for cell viability than L-glutamine limitation.

Nonessential amino acids

All media formulations contain the ten essential amino acids, as well as cysteine, glutamine, and tyrosine. Including other non-essential amino acids (alanine, asparagine, aspartic acid, glycine, glutamic acid, proline, and serine) in some media formulations reduces the metabolic burden on the cells allowing increased cellular proliferation.

Sodium pyruvate

Pyruvate is an intermediary organic acid metabolite in glycolysis and the first component of the Embden-Meyerhof pathway. It can readily pass in and out of the cell. Its addition to tissue culture medium provides both an energy source and a carbon skeleton for anabolic processes. Pyruvate may help maintain certain specialized cells, in clonal selection, in reducing the serum concentration of the medium, and in reducing fluorescent light-induced phototoxicity. Cellular metabolism of pyruvate produces carbon dioxide which is given off into the atmosphere and becomes bicarbonate in the medium.

Phenol red

Phenol red is used to monitor the pH of media. During cell growth, the medium changes color as it changes pH in response to metabolites released by the cells. At low pH levels, phenol red turns the medium yellow, while at higher pH levels, it turns the medium purple. For most tissue culture work (pH 7.4), the medium should be bright red. Unfortunately, phenol red can mimic the action of some steroid hormones, particularly estrogen. For studies with estrogen-sensitive cells, such as mammary tissue, phenol red cannot be used. Phenol red is also frequently omitted from studies with flow cytometry as its color interferes with detection.

Osmolality

The osmolality of cell culture media for most vertebrate cells is kept within a narrow range, from 260 mOsm/kg to 320 mOsm/

kg, even though most established cell lines will tolerate a rather large variation in osmotic pressure. In contrast, the osmolality requirements for some invertebrate cell lines fall outside of this range. For example, the snail embryo requires medium of about 155 mOsm/kg, while some insect cells prefer 360 mOsm/kg to 375 mOsm/kg. Most commercially available liquid media report osmolality and it is advisable to check the osmolality of any medium after adding of saline solutions, drugs, hormones dissolved in an acid or base solution, or large volumes of buffers (e.g., HEPES).

Media supplements

The complete growth media recommended for some cell lines requires the additional components not available in the basic media and serum. These components include hormones, growth factors, and signaling substances that sustain proliferation and maintain normal cell metabolism. Supplements are usually prepared as 100× (or higher) stock solutions in serum-free medium. Some supplements may need to be dissolved in a solvent prior to subsequent dilution in serum-free medium to the stock concentration. Stock concentrations should be aliquoted into small volumes and stored at an appropriate temperature; most stock concentrations can be stored at $-80\ ℃$.

The addition of supplements can change the final osmolality of the complete growth medium, potentially having a negative effect on the growth of cells in culture. It is best to recheck the osmolality of the complete growth medium after small volumes of supplement stock solutions are added.

Antibiotics and antimycotics

Antibiotics or antimycotic agents are added to cell culture media as a prophylactic to prevent contamination, as a cure after contamination is found, to induce expression of recombinant proteins, or to maintain selective pressure on transfected cells. Routine use of antibiotics or antimycotics for cell culture is not recommended unless they are specifically required, such as G418 for maintaining selective pressure on transfected cells. Antibiotics can mask contamination by mycoplasma and resistant bacteria. Further, they can interfere with the metabolism of sensitive cells. It is best to avoid antimycotics as they can be toxic to many cell lines. While cell lines can be cured of microbial contamination with antibiotics and/or antimycotics, this is not recommend unless the cell line is irreplaceable; the process is lengthy and there is no guarantee that contamination will be eliminated. Even if the contamination is eliminated, there is no way to ensure that the resulting cell line will have the same characteristics as the initial one due to the stress of the treatment. It is best to discard the cell line and start over with new stocks. Mycoplasma contamination is particularly difficult to eliminate. In some cases, antibiotic use for short periods of time can serve as a valuable prophylactic.

Animal sera

Sera serve as a source for amino acids, proteins, vitamins (particularly fat-soluble vitamins, such as A, D, E, and K), carbohydrates, lipids, hormones, growth factors, minerals, and trace elements. Additionally, serum buffers the culture medium, inactivates proteolytic enzymes, increases medium viscosity (which reduces shear stress during pipetting or stirring), and conditions the growth surface of the culture vessel. The exact composition is unknown and varies from lot-to-lot, although lot-to-lot consistency has improved in recent years.

Sera from fetal or calf bovine sources are commonly used to support the growth of cells in culture. Fetal serum is a rich source of growth factors appropriate for cell cloning and growing fastidious cells. Calf serum, because of its lower growth-promoting properties, is used in contact-inhibition studies with NIH/3T3 cells. In contrast to fetal or calf sera, horse serum is collected from a closed herd of adult animals ensuring lot-to-lot consistency. Horse serum is less likely to carry the contaminants found in bovine sera, such as viruses, and less likely to metabolize polyamines which may be mitogenic for some cells. Horse and bovine calf sera are less expensive and more readily available than fetal bovine serum. Unfortunately, naturally derived products from bovine sources may contain adventitious viruses such as bovine viral diarrhea virus (BVDV), bovine parvovirus, bovine adenovirus, and blue tongue virus. All reputable suppliers test their products for infectious virus by several methods, including fluorescent antibody, cytopathic effect, and hemadsorption. These products are also screened for the standard microbial contaminants such as bacteria, fungi, and mycoplasma. Bovine-derived products may also contain the agent responsible for bovine spongiform encephalopathy (BSE).

Storage

Store sera at $-20\,°C$ or colder for storage over 30 days. Sera are routinely stored at below $-70\,°C$. Do not store sera at temperatures above $-20\,°C$ for any length of time. Avoid repeated freeze-thaws by dispensing and storing in small aliquots.

Thawing

The following procedure is used to thaw serum:

(1) Place frozen serum in a refrigerator at 2 ℃ to 8 ℃ overnight.

(2) Put the bottle in a 37 ℃ water bath and gently agitate from time to time to mix the solutes that tend to concentrate at the bottom of the bottle.

Note: *Do not keep the serum at 37 ℃ any longer than is necessary to thaw it, and do not thaw the serum at higher temperatures. Thawing serum in a bath above 40 ℃ without mixing may lead to the formation of a precipitate inside the bottle.*

Turbidity and precipitates

All sera may retain some fibrinogen. External factors may initiate the conversion of fibrinogen to fibrin, leading to flocculent material or turbidity after serum is thawed. This material does not alter the serum's performance. If the presence of flocculent material or turbidity is a concern, it can be removed by filtration through a 0.45 μm filter.

A precipitate can form in serum when incubated at 37 ℃ or higher for prolonged periods of time and may be mistaken for microbial contamination. This precipitate may include crystals of calcium phosphate, but does not alter the performance of the serum as a supplement for cell culture. Heat inactivation of sera can also cause precipitates to form.

Heat inactivation

Heat inactivation is usually unnecessary and can be detrimental to the growth of some cells. It reduces or destroys the growth factors in the serum. Heat inactivation was originally performed to inactivate complement (a group of proteins present in sera that are part of the immune response) and to destroy mycoplasma contaminants. Today, mycoplasma contamination, if any, is removed by filtration. Removal of complement is usually unnecessary, but can be important when preparing or assaying viruses or in cytotoxicity tests.

Culture vessels and surfaces

Culture vessels provide a contamination barrier to protect cultures from the external environment while maintaining the proper internal environment. For anchorage-dependent cells, the vessels provide a suitable and consistent substrate for cell attachment. Other characteristics of vessels include easy access to the cultures and optically clear viewing surfaces.

Originally, all culture vessels were made of glass. Drawbacks of using glass include the heavy weight, expense, labor-intensive cleaning, and poor microscopic viewing compared to plastic. By the 1960s, surface treatment techniques were developed for polystyrene, allowing plastic vessels to replace glass for most cell culture applications.

Selecting the right vessel

First, match the characteristics of the cells to be grown with the characteristics of the different culturing systems. There are three basic types of cell cultures:

• Anchorage dependent cells, which must attach to a surface to grow (for example, human diploid fibroblasts).

• Anchorage independent cells, which grow in suspension (most blood-derived cell cultures).

• Cells that can grow either attached or in suspension (many transformed cell lines).

Understand the growth requirements of the cultures to help select the best culture system.

Stationary monolayer cultures are grown in undisturbed flasks, dishes, and multiwell plates. These are the easiest culture systems to use and require the least amount of equipment. However, these systems are very labor intensive for producing large quantities of cells.

Stationary suspension cultures are grown without agitation in untreated dishes and flasks. These are best for growing small volumes of anchorage-independent cells that grow poorly in traditional stirred suspension cultures.

Moving suspension cultures are grown in mechanically stirred vessels (spinner flasks), bioreactors, or fermenters. These systems are the most economical in terms of space, labor, and media; as a result, stirred suspension cultures are usually the method of choice for producing large volumes of cells, both in the lab and in industry.

Next, decide whether the cells will be grown as an open system or as a closed system.

Open-system plastic dishes are less expensive than closed-system flasks, but require more expensive incubators to regulate the CO_2 and humidity in the atmosphere. Closed culture vessels and surface systems provide additional protection against contamination and have simpler incubator requirements. All dishes and multiwell plates are open systems. All other culture vessels can be used in either mode by leaving caps loose for an open system or tightened for a closed system. The plastic walls of culture vessels are slightly permeable to carbon dioxide and oxygen, permitting a very small amount of gas exchange.

This is not a problem in most culture applications but may interfere with anoxia experiments or long-term media storage. Caps that allow gas exchange when the cap is fully tightened are also available to reduce

opportunities for flask spills and contamination in open systems.

The last step is matching the desired cell yield with an appropriately sized culture vessel. For monolayer cultures, the yield is limited by the area of treated growth surface. A typical yield for confluent continuous mammalian cell lines is approximately 0.5×10^5 cells/cm^2 to 1×10^5 cells/cm^2 of treated surface. For suspension cultures, the total cell yield is determined by the working volume of the vessel. In stirred systems, cell concentrations can easily reach between 1×10^6 cells/mL and 2×10^6 cells/mL of medium. However, the exact yields need to be determined empirically for each cell line.

Aseptic technique for cell culture

Bacteria can be isolated from nearly any surface, including animate objects and human skin. Fungals spores and bits of vegetative hyphae can drift into a laboratory from air conditioning ducts and open doors. Mycoplasma infections most frequently originate from improperly sterilized media or serum. Inherent with successful manipulation of cell cultures is the basic understanding that everything that comes into contact with the cells must be sterile or non-contaminating. This includes media, glassware, and instruments, as well as the environment to which the cultures are briefly exposed during transfer procedures. Because cleaning up a contaminated culture is too frequently a disheartening and unsuccessful experience, the best strategy is to prevent microbial contamination from occurring in the first place.

Aseptic technique

One basic concern for successful aseptic technique is personal hygiene. The human skin harbors a naturally occurring and vigorous population of bacterial and fungal inhabitants that shed microscopically and ubiquitously. Most unfortunately for cell culture work, cell culture media and incubation conditions provide an ideal growth environment for these potential microbial contaminants. Ideally, all aseptic work should be conducted in a laminar flow cabinet. However, work space preparation is essentially the same for working at the bench. Flame sterilization is used as a direct and localized means of decontamination in aseptic work at the open bench. It is most often used (1) to eliminate potential contaminants from exposed openings of media bottles, culture flasks, or test tubes during transfers; (2) to sterilize small instruments, such as forceps; or (3) to sterilize wire inoculating loops and needles before and after transfers. When possible, flame sterilization should be avoided in laminar flow environments because the turbulence generated by the flame can significantly disturb the sterile air stream. Infrared electric sterilizer is appropriate if necessary (Fig. 7-4).

Fig. 7-4　Infrared electric sterilizer

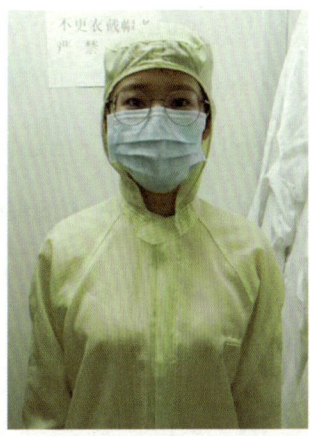

Fig. 7-5　Cuffed laboratory coat for cell culture

Take personal precautions

(1) Prior to aseptic manipulations, tie long hair back behind head. Vigorously scrub hands and arms for at least 2 min with antibacterial soap, because loosely adhering skin flora can easily dislodge and potentially fall into sterile containers.

For nonhazardous sterile applications, wear clean, cuffed laboratory coats and latex gloves (Fig. 7-5). Greater stringencies may be necessary depending upon the regulatory requirements of the laboratory. Front-closing laboratory coats are not recommended for work with hazardous biological agents. Safety glasses should be worn when manipulating biological agents outside of a biosafety cabinet.

(2) Frequently disinfect gloved hands with 75% ethanol while doing aseptic work. Although the gloves may have initially been sterile when first worn, they have surely contacted many nonsterile items while in use. Note that 75% ethanol may not be appropriate for latex glove disinfection when working with cultures containing animal viruses, as studies have shown that ethanol increases latex permeability, reducing protection for the wearer in the event of exposure. In this case, quarternary ammonium compounds are more appropriate for disinfection.

(3) Thoroughly wash hands after removing protective gloves.

Prepare and maintain the work area

(1) Perform all aseptic work in a clean work space, free from contaminating air currents and drafts. For optimal environmental control, work in a laminar flow cabinet.

(2) Clear the work space of all extraneous items that are not directly required for the aseptic operation being performed.

(3) Wipe down the work surface before and after use with 75% ethanol or other appropriate disinfectant (Fig. 7-6).

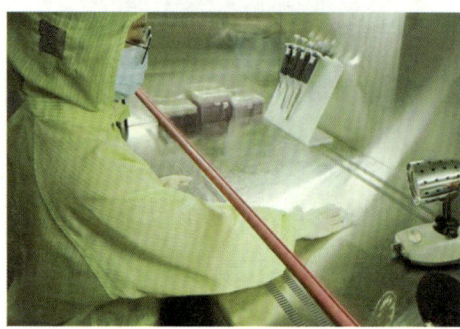

Fig. 7-6 Clean the bench with 75% ethanol or other appropriate disinfectant

(4) When feasible, wipe down items with disinfectant as they are introduced into the clean work space. Arrange the necessary items in the work space in a logical pattern from clean to dirty to avoid passing contaminated material (e.g., a pipet used to transfer cultures) over clean items (e.g., flasks of sterile media).

(5) Immediately dispose any small contaminated items into a discard pan.

(6) When the aseptic task has been completed, promptly remove any larger contaminated items or other material to be disposed (e.g., old culture material, spent media, waste containers) from the work space and place it in designated bags or pans for autoclaving.

(7) For a right-handed person, hold the vessel in the left hand at a $-45°$ angle (or titled as much as possible without spilling the contents) and gently remove its closure. Do not permit any part of the closure that directly comes in contact with the contents of the vessel to touch any contaminating object (e.g., hands or work bench). Holding the vessel off the vertical axis while opening will prevent any airborne particulates from entering the container.

(8) Slowly pass the opening of the vessel over the top of (rather than through) a burner flame to burn off any contaminants. Be careful when flaming containers of infectious material. Any liquid lodged in the threads of a screw cap container will spatter as it is heated. Aerosols thus formed this way may actually disseminate entrapped biological agents before the heat of the flame able to inactivate them.

(9) While still holding the vessel at a slant, use a sterile pipet and pipettor to slowly add or remove aliquots to avoid aerosol formation.

(10) Flame-sterilize again, allow the container to cool slightly, and carefully recap the vessel.

Section 1 Subculturing Cells

Background Reading

Anchorage-dependent cell lines growing in monolayers need to be subcultured at regular intervals to maintain them in exponential growth. When the cells are near the end of exponential growth (roughly 70% to 90% confluent), they are ready to be subcultured.

Subculturing monolayers involves breaking of both intercellular and intracellular cell-to-surface bonds. For some loosely attached cells, a sharp blow with the palm of your hand against the side of the flask can dislodge them. Many require their protein attachment bonds to be digested by proteolytic enzymes such as trypsin-EDTA. After the cells have been dissociated and dispersed into a single-cell suspension, they are diluted to the appropriate concentration and transferred to fresh culture vessels with the appropriate growth medium where they will reattach, grow, and divide. The procedure below is appropriate for most adherent cell lines. However, since every cell line is unique, incubation times, and temperature, number of washes, or the solution formulations may vary. In all cases, continuously observe the cells with a microscope during the dissociation process to prevent cellular damage caused by the dissociation solution. The amounts used in this procedure are appropriate for a 6 cm dish. Adjust volumes as appropriate for different sized vessels.

Objectives

- Cell subculturing
- Cell counting

Materials

Reagents: Basic medium (DMEM); Fetal bovine Serum (FBS); 0.25% Trypsin-EDTA (1 × TE); PBS ($Ca^{2+}-$, $Mg^{2+}-$ free); Trypan blue solution.

Suppliers: Petri-dish; Pipette; Centrifuge tubes; Microcentrifuge tubes.

Equipments: Biological safety cabinet; Cell counter; Inverted microscope; CO_2 incubator; Centrifuge; Water bath.

Activity Protocol

Part 1. For monolayer cells

(1) Bring the growth medium, trypsin-EDTA solution, balanced salt solution [Phosphate Buffered Saline (PBS) without

calcium or magnesium], and FBS to the appropriate temperature for the cell line. In most cases, this is the temperature used to grow the cells (usually 37 ℃). For some sensitive cells, the trypsin-EDTA solution may need to be used at room temperature or 4 ℃.

(2) Remove and discard the cell culture medium from the dish by vacuum aspirator (Fig. 7-7).

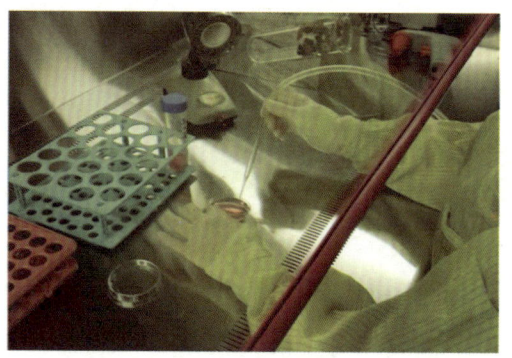

Fig. 7-7   Remove the cell culture medium

(3) Rinse the cell monolayer with 1 × PBS (1 to 2 mL) without calcium or magnesium and then remove the PBS.

(4) Add 0.5 mL of the trypsin-EDTA (0.25% TE) solution to a 6 cm dish and incubate at 37 ℃. Monitor the progress of cell dissociation by microscopy.

(5) Once the cells appear to be detached (3 to 10 min for most cell lines; they will appear rounded and refractile under the microscope, Fig. 7-8), add an equal volume of serum or 1 to 2 mL of complete growth medium with a pipette to the cell suspension to inactivate the trypsin.

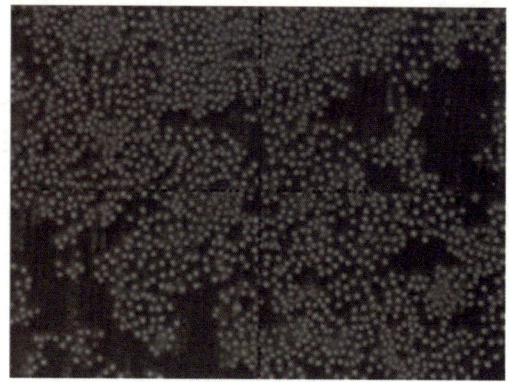

Fig. 7-8   Cells detached from the vessel

(6) Gently wash any remaining cells from the growth surface of the dish. Check the cells with the microscope to ensure that most (> 95%) are single cells. If cell clusters are present, continue to disperse the cells with gentle pipetting.

(7) Transfer the cells to a centrifuge tube and centrifuge at 900 rpm for 3-5 min.

(8) Remove and discard the supernatant from the tube by vacuum aspirator.

(9) Add the appropriate volume of fresh culture medium to the cell pellet and resuspend cells with gentle pipetting.

(10) Count the cells in suspension and determine their viability.

◆ Mix 10 μL of cell solution with 10 μL of trypan blue solution, stand the mixture for 3 min.

Note: *The mixture must be loaded into counting slide within 5 min; viable cells that are exposed to trypan blue dye for an extended period may start incorporating the dye, affecting the accuracy of the cell count.*

◆ Pipete 10 μL of mixture into the outer opening of either chamber of the counting slide (Fig. 7-9A).

Note: *When loading the sample, place the pipet tips at a 45° angle at the bottom of sample loading area. Care should be taken to avoid visible bubble formation or back splatter.*

◆ Insert the counting slide into the slide slot of the cell counter. Make sure that the slide is completely inside the slide slot (Fig. 7-9B).

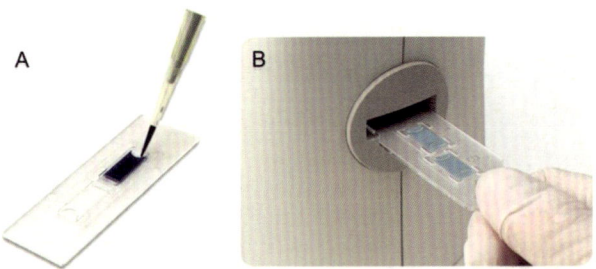

Fig. 7-9 Cell counting by automatical cell counter

◆ The cell counter automatically detects the presence of the slide and initiates the count.

◆ Calculate viable cell number per mL.

(11) Add a suitable amount of cell suspension to a complete medium in a centrifuge tube and gently blow after blending.

(12) Transfer the cells onto a new petri-dish.

Note: *Do not add a concentrated cell suspension to an empty culture vessel as this can result in uneven cell attachment and growth.*

(13) Place the cells back into the incubator.

(14) Examine the culture the following day to ensure the cells have reattached and are actively growing.

(15) Change the medium as needed. For most actively growing cultures, two to three times per week is typical.

Part 2. For suspension cells

Most primary cultures, finite cell lines, and continuous cell lines are anchorage-dependent and thus grow in monolayers attached to a surface. Other cells, particularly those derived from hematopoietic tissues or certain tumor tissues, are anchorage-independent and grow in suspension. Cell propagation in suspension has several advantages over propagation in a monolayer. Subculturing is a simple matter of dilution. There is little or no growth lag after splitting a suspension cult-

ure as there is with a monolayer culture because the cells experience none of the trauma associated with proteolytic enzyme dispersal. Suspension cultures require less lab space per cell yield, and scale-up is straightforward. Cells can be propagated in bioreactors similar to fermenters used for yeast or bacteria cultures.

◆ Bring the growth medium, 1 × PBS (calcium and magnesium free), and serum to the appropriate temperature for the cell line.

◆ Collect the cells in a centrifuge tube and centrifuge at 900 rpm for 3-5 min.

◆ Remove and discard the supernatant from the tube by vacuum aspirator.

◆ Rinse the cells with 1 × PBS and centrifuge at 900 rpm for 3-5 min.

◆ Discard the PBS by vacuum aspirator.

◆ Add the appropriate volume of fresh culture medium to the cell pellet and resuspend cells with gentle pipetting.

◆ Count the cells in suspension and determine their viability.

◆ Add a suitable amount of cell suspension to a complete medium in a centrifuge tube and gently blow after blending.

(16) Transfer the cells to a new petri-dish.

(17) Place the cells back into the incubator.

(18) Examine the culture the following day to ensure the cells are actively growing.

(19) Change the medium as needed. For most actively growing cultures, two to three times per week is typical.

Information Box 7-1: Cell viability

Viability assays measure the number of viable cells in a population. When combined with the total number of cells, the number of viable cells provides an accurate indication of the health of the cell culture. The most common and rapid methods rely on the integrity of the cell membrane as an indicator of cell viability.

Trypan blue dye is actively excluded by viable cells but is taken up and retained by dead cells, which lack an intact membrane. Nonviable cells will be stained dark blue. Cell viability is expressed as a percentage of the number of unstained or viable cells per the total number of cells.

Information Box 7-2: Hemocytometer

As hemocytometer is excellent for determining cell viability, but is not precise in determining cell number due to the relatively low number of cells actually counted. An automated counter will generate the most reliable data, particularly when used in combination with the viability data from a hemocytometer.

(1) Clean, thoroughly dry, and assemble the hemocytometer with the cover slip (Fig. 7-10A).

(2) Transfer a small amount of cell suspension (usually 10 μL) to the edge of each of the two counting chambers. Allow the cell suspension to be drawn into the counting chamber (Fig. 7-10B).

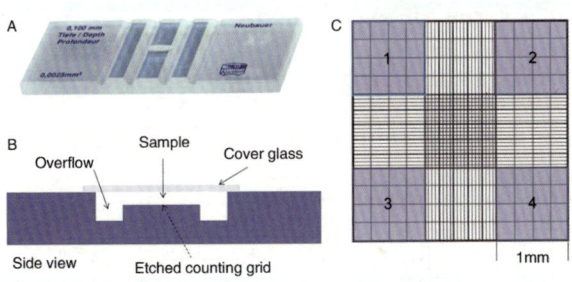

Fig. 7-10 Hemocytometer

(3) Place the hemocytometer under an inverted microscope and view the cells at 100 × magnification.

(4) Focus on the quadrants labeled 1, 2, 3, and 4 (Fig. 7-10C). Record the number of cells in each section. Average the number of cells, and multiply by the dilution factor. Any dilution of the sample after it was removed from the cell suspension, such as using vital stain, needs to be included in the calculation. For example, if the four counts are 60, 66, 69, and 75, the concentration would be 67.5×10^4 cells/mL for the sample that was loaded into the hemocytometer. For best results, adjust the concentration of the suspension so that 50 to 100 cells are in each of the four sections.

Part 3. Examining cultures

Observe the morphology and viability of cultures regularly and carefully. Additionally, it is important to examine the medium in the vessel for macroscopic evidence of microbial contamination. This includes unusual pH shifts (yellow or purple color of the phenol red), turbidity, or particles. Also, look for small fungal colonies that float at the medium-air interface. Specifically check around the edges of the vessel as these may not be readily visible through the microscope.

Using an inverted microscope at low magnification (40 ×), check the medium for evidence of microbial contamination and the morphology of the cells. Bacterial contamination will appear as small, shimmering black dots within the spaces between cells. Yeast contamination will appear as rounded or budding particles, while fungi will have thin filamentous mycelia. Most adherent cells should be attached firmly to the surface. In some cases, healthy cells will round up and detach somewhat during mitosis and appear very refractile. Following mitosis, they will reattach. Some of these will float free if the culture vessel is physically disturbed. In contrast, dead cells often round up and detach from the monolayer, appearing smaller and darker (not refractile) than healthy cells.

In addition to daily examinations, periodically test a sample of the culture for the presence of fungi, bacteria, and mycoplasma.

Section 2　Cryopreservation

Background Reading

Most cell cultures can be stored for many years, at temperatures below −170℃ (cryopreservation). The many advantages of cryopreservation far outweigh the required investment in equipment and reagents. The advantages include the following:

＊Generation of backup stocks to protect against loss of the cell line caused by equipment failures or contamination by microorganisms or other cell lines.

＊Elimination of the time, energy, and materials required to maintain cultures not in active use.

＊Preservation of cells with finite population doublings (that will ultimately senesce).

＊Insurance against phenotypic drift in the culture due to genetic instability and/or selective pressure.

＊Creating a standard cell tine for use in a series of experiments.

As the cell suspension is cooled below the freezing point, ice crystals form and solute concentration increases in the suspension. Intracellular ice can be minimized if water within the cell is allowed to escape through osmosis during the cooling process. A slow cooling rate, generally −1℃ per minute, facilitates this process. However, as the cells lose water, they shrink in size and will quickly lose viability if they go below a minimum volume. The addition of cryoprotectants such as glycerol or dimethylsulfoxide (DMSO) can mitigate these effects.

The standard procedure for cryopreservation is to freeze cells slowly in medium including a cryoprotectant until they reach a temperature below -70℃ in medium that includes a cryoprotectant. Vials are then transferred to a liquid-nitrogen freezer to maintain them at temperatures below -170℃. Recovering cryopreserved cells is straightforward: cells are thawed rapidly in a water bath at 37℃, removed from the freeze medium by gentle centrifugation and/or diluted with growth medium, and seeded in a culture vessel in complete growth medium.

Numerous factors affect the viability of recovered cells. Modify the procedure for each cell line to attain optimal cell viability upon recovery. Some critical parameters for optimization include the composition of the freeze medium, the growth phase of the culture at the time of freezing, the stage of the cell in the cell cycle, and the number and concentration of cells in the freezing solution.

Freeze Medium

Glycerol and DMSO at 5%-10% are the most common cryoprotectants. While DMSO can be toxic to cells, it penetrates much faster than glycerol and yields more reproducible results. Unfortunately, DMSO can cause some cells to differentiate (e.g., HL-60 promyeloblast cells) and may be too toxic for other cells (e.g., HBE4-E6/E7 lung epithelial cells). Glycerol should be used in these instances. Glycerol can be sterilized by autoclaving whereas DMSO must be sterilized by filtration.

Use only reagent-grade (or better, such as cell culture-grade) DMSO or glycerol. Store both in aliquots protected from light. For cells grown in serum-free medium, adding 50% conditioned medium (serum-free medium in which the cells were grown for 24 h) to both the cell freezing and the recovery media may improve recovery and survival. The addition of 10%-20% cell culture-grade bovine serum albumin to serum-free freezing medium may also increase postfreeze survival.

Equipment

Cryopreservation vials

There are two materials to choose for cryopreservation vials: plastic or glass. Plastic vials are used to store of distribution stocks. Plastic vials come in two varieties: those with an internal thread and silicone gasket and those with an external thread. The internal-thread version was the first commercially available kind but has some disadvantages over the external-thread version. For example, while the silicone gasket provides an excellent seal, it must be tightened just right; too tight or too loose and the vial will leak. Glass vials are more difficult to work with; they need to be sterilized before use, they do not come with labels (information is imprinted into the glass), they need to be sealed with a hot flame, and they can be difficult to open. However, they are preferred for long-term storage (many years) of valuable cultures and are considered failsafe once properly sealed.

Controlled-rate freezing boxes

There are several means to achieve a cooling rate of $-1\ ℃$ per minute. A relative low-cost approach is to place the cryopreservation vials into an insulated chamber and then cool it for 24 h in a mechanical freezer at $-70\ ℃$ or lower. There are several commercially available freezing boxes (Fig. 7-11) that achieve a cooling rate very close to the ideal $-1\ ℃$ per minute. Programmed cooling boxes are used for a variety of cell types, including stem cells, primary cells, and cell lines. The programmed cooling box does not require any liquid or additives and only needs to be put into an ultra-low temperature freezer to ensure that the cooling rate of the samples in the cooling box is $-1\ ℃$ per minute.

Fig. 7-11　Programmed freezing boxes

Liquid nitrogen freezer storage

The ultra-low temperatures (below $-170\ ℃$) required for long-term storage can be maintained using specialized electric freezers or, more commonly using liquid nitrogen freezers (Fig. 7-12). There are two basic types of liquid nitrogen storage systems: immersing vials in the liquid-phase or holding vials in the vaporphase above the liquid. The liquid-phase system holds more nitrogen and thus requires less maintenance. However, there is always a chance that some liquid will enter improperly sealed vials, potentially causing them to explode when retrieved. For this reason vapor-phase systems are strongly recommended. Vapor-phase systems create a vertical temperature gradient within the container. The temperature in the liquid nitrogen at the bottom will be $-196\ ℃$, while the temperature at the top will vary depending upon the amount of liquid nitrogen at the bottom and the amount of time the container is opened. To ensure safe storage of cells, keep enough liquid nitrogen in the container to maintain the temperature at the top at $-130\ ℃$ or colder. All storage systems should be equipped with temperature alarms.

Fig. 7-12　Liquid nitrogen freezer

✜ Objective

❖ Cryopreservation of monolayer cells

✜ Materials

Reagents: Basic medium (DMEM); Fetal bovine Serum (FBS); 0.25% Trypsin-EDTA (1 × TE); PBS (Ca^{2+}-, Mg^{2+}-free); Trypan blue solution; DMSO

Suppliers: Petri-dish; Freezing vial; Programmed freezing box; Pipette; Centrifuge tubes; Microcentrifuge tubes

Equipment: Biological safety cabinet; Cell counter; Inverted microscope; CO_2 incubator; Centrifuge; Water bath; −80 ℃ refrigerator; Liquid nitrogen tank

✜ Activity protocol

(1) Check your cell culture for contamination of bacteria, fungi, mycoplasma, and viruses immediately before cryopreservation.

Note: *Cells with any contamination should not be stored.*

(2) Prepare a freeze medium consisting of complete growth medium and 10% DMSO (or glycerol). Do not add undiluted DMSO to a cell suspension as DMSO dissolution in aqueous is an exothermic reaction, and may harm the cells in suspension.

Note: *Be careful while handling any DMSO solution as it will rapidly penetrate intact skin and may carry toxic contaminants along with it.*

(3) Collect cells using gentle centrifugation (5 min at 900 rpm) and resuspend them in the freeze medium at a concentration of 1×10^5 to 5×10^5 viable cells/mL. Continue maintaining the cells in culture until viability of the recovered cells is confirmed.

(4) Label the appropriate number of vials with the name of the cell line and the date (Fig. 7-13). Next, add 1 to 1.5 mL of the cell suspension to each vial (depending on the volume of the vial) and seal.

(5) Allow cells to equilibrate in the freeze medium at room temperature for a minimum of 15 min but no longer than 40 min.

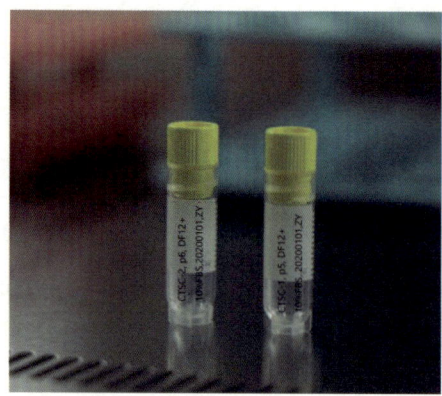

Fig. 7-13　Label the freezer vials

Note: *This time is usually taken up by dispensing aliquots of the cell suspension into the vials. After 40 min, cell viability may decline due to the DMSO.*

(6) Place the vials into a programmed freezing box and place the box in a mechanical freezer at −70 ℃ (or colder) for at least 24 h.

(7) Quickly transfer the vials to a liquid nitrogen or −170 ℃ freezer. Frozen material will warm up at a rate of 10 ℃ per minute and cells will deteriorate rapidly if warmed to temperatures above −50 ℃.

(8) Record the location and details of the freezer.

(9) After 24 h at −170 ℃, remove one vial, revive the cells in culture medium, and determine their viability and sterility.

Section 3　Recovery of Cryopreserved Cells

Background Reading

The cell solution in the frozen vial needs to be warmed as rapidly as possible and then immediately combined with complete culture medium and seeded into an appropriate dish or flask. Most cells show a decline in viability at 24 h post-thaw because of apoptosis induced by the stress of the cryopreservation process. After this time point, cells begin to recover and enter the exponential growth phase.

Prepare a culture vessel so that it contains the appropriate culture medium equilibrated for temperature and pH.

Objective

❖ Recovering monolayer cells

Materials

Reagents: Basic medium (DMEM); Fetal bovine Serum (FBS)

Suppliers: Petri-dish; Pipette; Centrifuge tubes

Equipment: Inverted microscope; CO_2

incubator; Centrifuge; Water bath

⚙ Activity protocol

(1) Preparation before recovery: Turn on the water bath to 37 ℃. Take two centrifuge tubes for cell collection and prepare complete culture medium.

(2) Remove the vial of cells from the liquid nitrogen freezer and thaw by gentle agitation in a 37 ℃ water bath. Thaw rapidly until ice crystals have just melted (Fig. 7-14).

Fig. 7-14   Thaw the cells by gentle agitation in a 37 ℃ water bath

Note: When taking the vial from the liquid nitrogen tank, wear anti-freezing gloves and protective goggles. Do not touch the tube body directly. Hold the top of the vial carefully. The liquid level of the water bath should not exceed the cover of the vial to prevent contamination. The recovery process must be rapid; the thawing time of frozen cells in a 37 ℃ water bath should be controlled between 1 min and not more than 3 min.

(3) Remove the vial from the water bath and decontaminate it by dipping in or spraying with 75% ethanol.

(4) Unscrew the top of the vial and transfer the contents to a sterile centrifuge tube containing complete medium.

(5) Remove the cryoprotectant by gentle centrifugation (5 min at 900 rpm).

(6) Discard the supernatant and resuspend the cells in 1 mL or 2 mL of complete growth medium. Pipette gently to loosen the cell pellet.

(7) Transfer the cell suspension to medium in the culture vessel and mix thoroughly.

(8) Examine the cultures after 24 h and subculture as needed.

Section 4   Know More

Serum-free Culture

Although serum-containing culture media promote cell growth, the use of serums in culture media has many disadvantages, including high cost, composition specificity, batch-to-batch variability, and animal ethics. Cell culturing with serum involves working with a product of animal origin. The most important problem is that the specific components of serum cannot be determined. The complexity of serum leads to corresponding difficulties in the cytological analysis. In 1976, Professor Gordon H. Sato (Fig. 7-15) proposed the concept of serum-free culture and identified several key factors that could replace serum[1-2]. Since then, serum-free culture media have been widely studied and employed, especially with the rapid development of stem cell science and regenerative medicine. In recent years, the application of serum-free, xeno-free, chemically defined, non-animal origin, and protein-free cell culture medium has spread rapidly.

Fig. 7-15 Professor Gordon H. Sato
(1927-2017)

Definition

Serum-free cell culture avoids the use of animal serum by replacing the serum with appropriate nutrient and hormone formulations, thus eliminating the uncertainty of serum composition from cell culture. The composition of serum-free cell medium is a standard formulation, facilitating the industrial production process and ensuring good experimental reproducibility.

Main advantages

❖ Known material composition. Serum-free culture medium can make the analysis system clear, and improve the reproducibility and stability of the experiment. The analytical system is controllable and can accurately evaluate the function of cells, providing better control over the physiological response of cells when external conditions change.

❖ High quality of cell products. Downstream processes can be simplified, and purification of cell products can be facilitated.

❖ Appropriate growth factors can be selected for different types of cells.

❖ No animal ingredients. Most of the added proteins are recombinant proteins or other synthetic substances, lacking animal-derived substances and have great application prospects in biomedical and other fields.

References

[1] HAYASHI I, SATO G. Replacement of serum by hormones permits the growth of cells in a defined medium [J]. Nature, 1976, 259: 132-134.

[2] Barnes D, Sato G. Serum-free cell culture: A unifying approach [J]. Cell, 1981, 22: 649-655.

Postlab Focus Questions

1. How do you know when it is time to passage your cells?

2. If you want to transfer cells to another city far from your city, how can you do so?

(Written by Zhang Yan)

Chapter 8   Cell Transfection

Background Reading

Transfection commonly refers to the process of artificially introducing nucleic acids (DNA or RNA) into eukaryotic cells, or more specifically, into animal cells. The word transfection is a blend of trans- and infection. The original meaning of transfection was "infection by transformation," i.e., introduction of DNA or RNA from a prokaryote-infecting virus or bacteriophage into cells, resulting in an infection. The term transformation had another sense in animal cell biology (a genetic change allowing long-term propagation in culture or acquisition of properties typical of cancer cells), hence the term transfection was used, and this technique is often used for non-viral methods in eukaryotic cells.

Transfection is a powerful tool for analyzing the functions of specific genes. Transfection deliberately introduces nucleic acids into cells, and typically involves opening transient pores or "holes" in the cell membrane of animal cells to allow the uptake of exogenous material. Genetic material (such as supercoiled plasmid DNA or siRNA constructs) or even proteins such as antibodies may be transfected. Introducing foreign nucleic acids using various chemical, biological, or physical methods can change the properties of the cell, allowing gene function and protein expression to be studied in the context of the cell.

Type of transfection

In transfection, the introduced nucleic acid may exist in the cells transiently, such that it is only expressed for a limited period of time and does not replicate; or it may be stable and integrate into the genome of the recipient, replicating when the host genome replicates. Based on these two situations, transfection strategies can be broadly classified into two general types: transient transfection and stable transfection.

Transient transfection

In transient transfection, the introduced nucleic acid exists in the cell for only a limited period of time and is not integrated into the genome. As such, transiently transfected genetic material is not passed from generation to generation during cell division, and can be lost because of environmental factors or diluted out during cell division. However, a high copy number of the transfected genetic material leads to high levels of protein expression while it is in the cell.

Depending on the construct used, transiently expressed transgenes can generally be detected for 1 to 7 days after transfection; however, transiently transfected cells are typically harvested 24 to 96 hours post-transfection. RNA or protein isolation for enzymatic assays or immunoassays may be required to analyze gene products. The optimal time interval for harvesting depends on the cell type, research goals, and specific expression characteristics of the introduced gene, as well as the time it takes for the reporter to reach steady state. However, within a few days, most of the foreign DNA is degraded by nucleases or diluted by cell division; after a week, its presence is no longer detected.

Transient transfection is most efficient when supercoiled plasmid DNA is used, presumably due to its more efficient uptake by the cell. siRNAs, miRNAs, mRNAs, and even proteins can be transiently transfected into cells, but as with plasmid DNA, these macromolecules need to be of high quality and purity. While transfected DNA is translocated into the nucleus for transcription, transfected RNA remains in the cytosol, where it is expressed within minutes after transfection (mRNA) or bound to mRNA to silence target gene expression (siRNA and miRNA).

Stable transfection

In stable transfection, foreign DNA is either integrated into the cellular genome or maintained as an episomal plasmid. Different from transient transfection, stable transfection allows the long-term maintenance of exogenous DNA in the transfected cell and its progeny. As such, stable transfection provides consistent expression of the introduced gene across multiple generations, which can be useful for producing recombinant proteins and analyzing downstream or long-term effects of exogenous DNA expression. However, a single or a few copies of the exogenous DNA are typically integrated into the genome of the stably transfected cell. For this reason, expression levels of stably transfected genes tend to be lower than those of transiently transfected genes (Fig. 8-1 and Table 8-1).

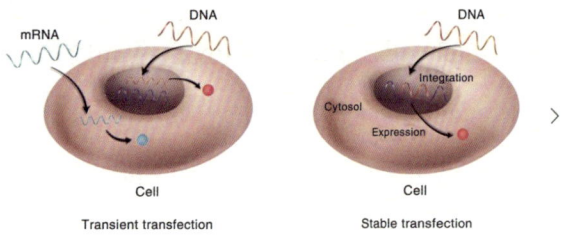

Fig. 8-1　Schematic diagrams of two different transfection methods

Stable integration of foreign DNA into the genome is a relatively rare event; therefore, successful stable transfection requires both effective DNA delivery and a way to select cells that have acquired the DNA. One of the most reliable ways to select cells stably expressing transfected DNA is to include a selectable marker in the DNA construct used for transfection and then apply the appropriate selective pressure to the cells after a short recovery period. Frequently used selectable markers include genes conferring resistance to various selection drugs or genes that compensate for an essential gene that is defective in the cell line being transfected. When cultured in selective medium, cells that were not transfected or were transiently transfected eventually die, and those that express the antibiotic resistance gene at sufficient levels or those that can compensate for the defect in the essential gene survive. Transient transfection is often used for studying the effects of short-term expression of genes or gene products, performing RNA interference (RNAi)-mediated gene silencing, or rapidly producing recombinant proteins on a small scale. In contrast, stable transfection is more useful when long-term gene expression is required or when transfected cells need to be used over the course of many experiments. Because integration of a DNA vector into the chromosome is a rare event, stable transfection of cells is more laborious and challenging, requiring selective screening and clonal isolation. As such, it is normally reserved for large-scale protein production, longer-term pharmacology studies, gene therapy, or research on the mechanisms of long-term genetic regulation.

Table 8-1 Comparison of the transient transfection and stable transfection

Transient Transfection	Stable Transfection
Transfected DNA is not integrated into the genome, but remains in the nucleus	Transfected DNA integrates into the genome
Transfected genetic material is not passed onto the progeny; genetic alteration is not permanent	Transfected genetic material is carried stably from generation to generation; genetic alteration is permanent
Does not require selection	Requires selective screening to isolate stable transfectants
Both DNA vectors and RNA can be used for transient transfection	Only DNA vectors can be used for stable transfection; RNA by itself cannot be stably introduced into cells
High copy number of transfected genetic material results in high levels of protein expression	Single or low copy number of stably integrated DNA results in lower levels of protein expression
Cells are typically harvested within 24-96 h of transfection	Requires 2-3 weeks of selection to isolate stably transfected colonies
Generally not suitable for studies using vectors with inducible promoters	Suitable for studies using vectors with inducible promoters

Different transfection methods

Transfection technologies available today can be broadly classified into three groups: chemical methods that use carrier molecules to neutralize or impart a positive charge to the negatively charged nucleic acids (Table 8-2); physical methods that directly deliver nucleic acids into the cytoplasm or the nucleus of the cell; and biological methods that rely on genetically engineered viruses to transfer non-viral genes into cells (also known as transduction, Table 8-3). However, no single method can be applied to all cell types and all experiments (Table 8-4). The ideal approach should be selected depending on different cell types and experimental needs, should have high transfection efficiency, low cell toxicity, minimal effects on normal physiology, and be easy to use and reproducible.

Table 8-2 Chemical gene delivery methods

Technology	Advantages	Disadvantages
Cationic lipid-mediated delivery	• Fast and easy protocols • Commercially available with reproducible results • High efficiency and expression performance • Applicable to a broad range of cell lines and high-throughput screens • Can be used for delivering DNA, RNA, and proteins • No size limitation on the packaged nucleic acid • Applicable to both transient and stable transfection • Can be used for *in vivo* delivery of nucleic acids	• Optimization may be necessary, some cell lines are sensitive to cationic lipids • Some cell lines are not readily transfected with cationic lipids • Presence of serum may interfere with complex formation and lower transfection efficiency • Absence of serum in the medium may increase cytotoxicity
Calcium phosphate co-precipitation	• Inexpensive and easily available • Applicable to both transient and stable transfection • High efficiency (cell line dependent)	• Requires careful preparation of reagents. $CaPO_4$ solutions are sensitive to changes in pH, temperature, and buffer salt concentrations • Reproducibility can be problematic • Cytoxicity, especially in primary cells • Does not work with RPMI due to high phosphate concentration of the medium • Not suited for *in vivo* gene transfer to whole animals
DEAE-dextran	• Relatively simple technique • Reproducible results • Inexpensive	• Chemical cytotoxicity in some cell types • Limited to transient transfection • Low transfection efficiency, especially in primary cells
Delivery by other cationic polymers (e.g., polybrene, PEI, dendrimers)	• Typically stable in serum and not temperature sensitive • High efficiency (cell line dependent) • Reproducible results	• Cytotoxicity in some cell types • Non-biodegradable (dendrimers) • Limited to transient transfection

Table 8-3 Biological gene delivery methods

Technology	Advantages	Disadvantages
Viral delivery	• Highest efficiency amongst gene delivery methods (80%-90% transduction efficiency in primary cells) • Works well with difficult to transfect cell types • Can be used for *in vivo* delivery of nucleic acids • Can be used to make stable cell lines (retroviral vectors) or for transient expression (adenoviral vectors)	• Cell lines to transfect must contain viral receptors • Limited insert size (~10 kb for most viral vectors versus; ~100 kb for non-viral vectors) • Technically challenging and time consuming to generate recombinant viruses • Present biosafety issues (activation of latent disease, immunogenic reactions, cytotoxicity, insertional mutagenesis, malignant transformation of cells)

Table 8-4 Physical gene delivery methods

Technology	Advantages	Disadvantages
Electroporation	• Simple principle • Reproducible results after optimization • No need for a vector • Less dependent on cell type and condition • Rapid transfection of large numbers of cells after optimization	• Requires a special instrument • Optimization of electrical pulse and field strength parameters required • Significantly more manipulation of cells required • High toxicity levels may be observed • High mortality rate requires large numbers of cells • Irreversibly damage the membrane and lyse the cells
Biolistic particle delivery (particle bombardment)	• Less dependent on cell type and condition and can be used for *in vivo* delivery of nucleic acids • Straightforward method with reliable results • No limitation to the size and/or number of genes that can be delivered • Primarily used for genetic vaccination and agricultural applications	• Requires an expensive instrument • Causes physical damage to samples • High mortality rate requires large numbers of cells • Preparation of microparticles is required • Relatively costly for research applications • Generally less efficient than electroporation or viral, or lipid-mediated delivery

(Continued)

Technology	Advantages	Disadvantages
Direct micro-injection	• Less dependent on cell type and condition • Allows single-cell transfection • Straightforward method with reliable results • No limitation to the size and/or number of genes that can be delivered • No need for a vector	• Requires an expensive instrument • Technically demanding and very labor-intensive (one cell at a time) • Often causes cell death
Laser-mediated transfection (photo-transfection)	• Can be used for delivering DNA, RNA, proteins, ions, dextrans, small molecules, and semiconductor nanocrystals • Can be applied to very small cells • Allows single-cell transfection or transfection of large number of cells at the same time • No need for a vector • High efficiency • Applicable to a broad range of cell lines	• Requires an expensive laser-microscope system • Requires cells to be attached • Technically demanding

Cationic lipid-mediated delivery

Cationic lipid-mediated transfection is one of the most popular methods to introduce foreign genetic material into cells. In this method, cationic lipids, such as 2,3-dioleyloxy-N-[2 (sperminecarboxamidio) ethyl]-N,N- dimethyl-l-propananminium trifluoroacetate (DOSPA) and the helper lipid dioleoyl phosphatidylethanolamine (DOPE) form liposomes, i.e., small, membrane-bound bodies that are somewhat similar to the structure of a cell and can fuse with the cell membrane to release DNA into the cell. For eukaryotic cells, transfection is better achieved using cationic liposomes because the cells are more sensitive. Cationic lipid-based reagents spontaneously form condensed nucleic acid-cationic lipid reagent complexes via electrostatic interactions between the negatively charged nucleic acid and the positively charged head group of the synthetic lipid reagent. These complexes are believed to be taken up by the cell through endocytosis and then released in the cytoplasm. The main advantages of lipofection are high efficiency, ability to transfect all types of nucleic acids in a wide range of cell types, ease of use, reproducibility, and low toxicity. In addition, this method is suitable for all transfection applications (transient, stable, co-transfection, reverse, sequential, or multiple transfections). High-throughput screening assays have also shown good efficiency in some *in vivo* models.

Mechanism of lipofectamine-mediated transfection

Specially designed cationic lipids, such as the Lipofectamine Transfection reagents, facilitate DNA or siRNA delivery into cells. The basic structure of cationic lipids consists of a positively charged head group and one or two hydrocarbon chains. The charged head group governs the interaction between the lipid and the phosphate backbone of the nucleic acid to facilitate DNA condensation. Often, cationic lipids are formulated with a neutral co-lipid or helper lipid, followed by extrusion or microfluidization to produce a unilamellar liposomal structure with a positive surface charge in water.

The positive surface charge of the liposomes mediates interaction between the nucleic acid and the cell membrane, allowing the liposome/nucleic acid transfection complex to fuse with the negatively charged cell membrane. The transfection complex is thought to enter the cell through endocytosis. Endocytosis is the process by which a localized region of the cell membrane uptakes the DNA/liposome complex by forming a membrane-bound intracellular vesicle. Once inside the cell, the complex must escape the endosomal pathway, diffuse through the cytoplasm, and enter the nucleus for gene expression. Cationic lipids are thought to facilitate transfection during the early steps of the process by mediating DNA condensation and DNA/cellular interactions (Fig. 8-2).

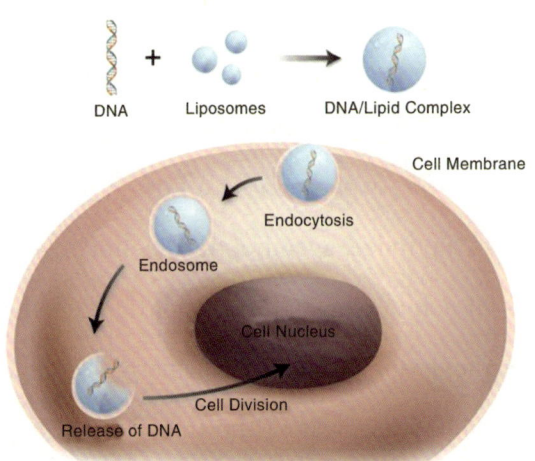

Fig. 8-2 Mechanism of transfection

Some problems associated with traditional transfection methods, such as calcium phosphate co-precipitation, DEAE-dextran, and polybrene include low DNA delivery efficiency, poor reproducibility, cell toxicity, and inconvenience. In contrast, cationic lipid reagent-mediated transfection yields high and previ-

ously unattainable transfection efficiencies in a wide variety of eukaryotic cells. It is simple to perform and ensures consistently reproducible results. Moreover, a number of cell lines normally resistant to transfection by other methods can be successfully transfected with cationic lipid reagents.

Factors influencing transfection efficiency

Successful transfection is influenced by many factors: the choice transfection method, health and viability of the cells, number of passages, degree of confluency, quality and quantity of the nucleic acid used, and the presence or absence of serum in the medium can all play a part in the outcome of your transfection experiment. While it is possible to optimize specific transfection conditions to achieve high transfection efficiencies, it is important to note that some cell death is inevitable regardless of the transfection method used.

Cell health and viability

The viability and general health of cells prior to transfection is an important source of transfection variability. In general, cells should be at least 90% viable prior to transfection and have sufficient time to recover from passaging. Subculturing cells at least 24 h before transfection is strongly recommend to ensure recovery from the subculture procedure and optimal physiological conditions for transfection. Cell cultures with immortalized cell lines evolve over months and years in the laboratory, resulting in changes in cell behavior with regard to transfection. Excessive passaging is likely to have detrimental effects on transfection efficiency and total transgene expression from the cell population as a whole. In general, cells that have undergone less than 30 passages after thawing a stock culture are recommended for use in transfection. Thawing a fresh vial of frozen cells and establishing low-passage cultures for transfection experiments allows recovery of transfection activity. For optimal reproducibility, aliquots of cells with a low passage number can be stored frozen and thawed as needed. Allow 3 or 4 passages after thawing a new vial of cells prior to transfection. Since contamination can drastically alter transfection results, cell cultures and media should be routinely tested for biological contamination, and contaminated cultures and media should never be used for transfection. If cells have been contaminated or their health is compromised in any way, they should be discarded and the culture reseeded from uncontaminated frozen stocks.

Cell confluency

For optimal transfection results, follow a routine subculturing procedure and passage cultures once or twice a week at a dilution that allows them to become nearly confluent before the next passage. Do not allow the cells to remain confluent for more than 24 h. The optimal cell density for transfection varies for different cell types, applications, and transfection technologies, and should be determined for every new cell line being transfected. Maintaining a standard seeding protocol from experiment to experiment reliably ensures optimal confluency at the time of transfection. With cationic lipid-mediated transfection, generally 70%-90% confluency is ideal for adherent cells. Make sure that the cells are not confluent or in stationary phase at the time of transfection because actively dividing cells take up foreign nucleic acid better than quiescent cells. Too high of a cell density can cause contact inhibition, resulting in poor uptake of nucleic acids and/or decreased expression of the transfected gene. However, too few cells in culture may result in poor growth without cell-to-cell contact. In such cases, increasing the number of cells in culture improves the transfection efficiency.

Culture media

Different types of cells have very specific medium, serum, and supplement requirements, and choosing the most suitable medium for the given cell type and transfection method plays a very important role in transfection experiments. Information for selecting the appropriate medium for a given cell type and transfection method is usually available in published literature, and may also be obtained from the source of the cells or cell banks. If there is no information available on the appropriate medium for your cell type, it must be determined empirically. It is important to use fresh media, especially if any components are unstable, because media missing key components and necessary supplements may harm cell growth. Some cell lines and primary cells may need special coating materials (e.g., poly-lysine, collagen, fibronectin, etc.) to attach to the culture plates and achieve optimal transfection results.

Serum

In general, the presence of serum in culture medium enhances transfection with DNA. However, when performing cationic lipid-mediated transfection, it is important to form DNA-lipid complexes in the absence of serum because some serum proteins interfere with complex formation. Note that the optimal amounts of cationic lipid reagent and DNA may change in the presence of serum; thus, transfection conditions should be optimized when using serum-containing transfection medium. When transfecting cells with RNA, we recommend performing the transfection procedure in the absence of serum to avoid possible RNases contamination. Most cells remain healthy for several hours in serum-free medium. The quality of serum can significantly affect cell growth and transfection results. Therefore, it is important to control for variability among different brands or even different lots of serum to obtain best results. After testing the serum on cells, keep using the same lot of serum to avoid variation in results.

Antibiotics

In general, antibiotics can be present in the medium for transient transfection. However, because cationic lipid reagents increase cell permeability, they may also increase the amount of antibiotics delivered into the cells, resulting in cytotoxicity and lower transfection efficiency. Therefore, avoid using antibiotics before and during cell transfection.

Type of molecule transfected

Plasmid DNA is the most commonly used vector for transfection. The topology (linear or supercoiled) and the size of the plasmid DNA vector influence transfection efficiency. Transient transfection is most efficient with supercoiled plasmid DNA. In stable transfection, linear DNA results in lower cellular DNA uptake relative to supercoiled DNA, but yields optimal integration of DNA into the host genome. Although other macromolecules, such as oligonucleotides, RNA, siRNA, and proteins can also be transfected into cells, conditions that work for plasmid DNA need to be optimized for other macromolecules.

Transfection method

There are a number of strategies for introducing nucleic acids into cells using various biological, chemical, and physical methods. However, not all of these methods can be applied to all types of cells and experimental applications. There is wide variation with respect to transfection efficiency, cell toxicity, effects on normal physiology, and gene expression levels. The ideal approach should be selected depending the cell type and experimental needs, and should have a high transfection efficiency, low cell toxicity, minimal effects on normal physiology, and be easy to use and reproducible.

Section 1 Transient Transfection

Objectives

- Transient transfection with Lipofectamine™ 3000 reagent
- Detection of transfection efficiency

Materials

Mammalian cells: Human embryonic kidney HEK 293T-EGFP cells (HEK 293T^{EGFP})

Reagents: DMEM culture medium; FBS; Penicillin-Streptomycin (P/S); Opti-MEM® I reduced serum medium; Lipofectamine™ 3000 Reagent; Plasmids

Suppliers: Petri-dish; microcentrifuge tube

Equipment: Biological safety cabinet; Electric pipette; Electric sterilizer; CO_2 incubator (37℃); Water bath; Inverted fluorescence microscope; Hemocytometer or cell counter

Activity protocol

Part 1. Transient transfection of HEK 293T^{EGFP} cells

(1) The day before transfection, seed 0.5×10^6-1.0×10^6 HEK 293T^{EGFP} cells in a 60 mm petri-dish with 4 mL DMEM culture medium (with 10% FBS and 1% P/S). Prepare three dishes for each experimental group.

(2) Incubate the cells at 37℃ in a CO_2 incubator until the cells are 60%-70% confluent. This will usually take 18-24 h, but the time will vary among cell type (optimal cell density may vary with cell types or application). Since transfection efficiency is sensitive to culture confluence, it is important to maintain a standard seeding protocol from experiment to experiment.

(3) For each 60 mm dish in a transfection, dilute 0 μg plasmid (negative control) or 6 μg target plasmids in 250 μL Opti-MEMR Medium (in a 1.5 mL microcentrifuge tube), then add 12 μL P3000™ Reagent to the tube. Mix gently and incubate at room temperature for 5 min.

(4) For each 60 mm dish in a transfection, dilute 15 μL Lipofectamine™ 3000 Reagent in 250 μL Opti-MEM Medium (in a 1.5 mL microcentrifuge tube). Mix gently and incubate at room temperature for 5 min.

(5) Combined diluted DNA with P3000™ Reagent (from step 3) and diluted Lipofectamine™ 3000 reagent (from step 4), mix gently, and incubate at room temperature for 10-15 min.

(6) During the incubation to form the DNA-liposome complex, take the cells from the CO_2 incubator, discard the culture medium, and wash the cells once with 2 mL Opti-

MEM Medium.

(7) For each transfection, add 0.5 mL of Opti-MEM Medium to the microcentrifuge tube containing the DNA-liposome complex. Mix gently and add the diluted complex solution to the rinsed cells (with 1 mL Opti-MEM Medium in each dish).

(8) Incubate the cells with the transfection complexes at 37 ℃ in a CO_2 incubator for 24-48 h.

Part 2. Observation of transient transfection results

(1) Carefully take the dishes with transfected cells from the CO_2 incubator and observe the fluorescent signal with an inverted fluorescence microscope.

(2) Select at least three views of a 60 mm dish and take photos of the bright field image and fluorescence image for each view.

◆ For a given view of a 60 mm dish with transfected cells, count the total cell number from the bright field image and the number of cells expressing fluorescent proteins.

◆ Calculate the transfection efficiency according to the following formula: fluorescent cells (sum of three views) / total cells (sum of three views) × 100%.

Section 2 Detecting Transfection Efficiency by Flow Cytometry

Background Reading

Flow cytometry (FCM) is a platform for rapid and quantitative analysis or sorting of a single row of cells or biological particles in a state of rapid linear flow. From the earliest prototype in 1934 to the formal introduction in 1969, FCM has developed rapidly in basic and applied life science research in the past 80 years.

FCM combines fluidics, optical components, electronics, and computer technologies to provide information about intracellular and extracellular cell characteristics. The latest FCMs are able to analyze several thousand particles every second, in "real time", and can actively separate and isolate particles with specified properties.

Data Analysis

FCM works with a computer system. A computer program controls the cytometer during data acquisition to:

∗ Select the parameters for measurement

∗ Select area, width, or height of different parameters

∗ Adjust the voltages on the photomultiplier tubes (PMTs)

∗ Adjust the gain settings on the amplifiers

∗ Select logarithmic or linear amplification

∗ Select and adjust the threshold (discriminator) settings

∗ Adjust color compensation settings

∗ Select histograms and cytograms for display

∗ Draw regions and set gates to be used during data acquisition

Gating

The data generated by FCM can be plotted in a single dimension, to produce a histogram, or in two-dimensional dot plots, or even in three dimensions. The regions on these plots can be sequentially separated based on fluorescence intensity by creating a series of subset extractions, termed "gates". Specific gating protocols exist for diagnostic and clinical purposes, especially in the field of hematology. The plots are often made on logarithmic scales. Because different fluorescent dyes have overlapping emission spectra, signals at the detectors must be electronically and computationally compensated. Data aquired using the FCM can be analyzed with software. After the data is collected, there is no need to stay connected to the FCM and analysis is typically performed on a separate computer. This is especially necessary in core facilities where usage of these machines is in high demand.

Objectives

❖ Learning the principle and application of FCM

❖ Operation of a BD Accuri™ C5 Flow Cytometer

❖ Detecting transfection efficiency by FCM

Materials

Mammalian cells: HEK293T^{EGFP} (Human embryonic kidney 293T) cells

Reagents: 0.25% Trypsin-EDTA; DMEM (with 10% FBS and 1% Penicillin-Streptomycin); 1 × PBS

Supplies: Petri-dish; Microcentrifuge tubes

Equipment: BD Accuri™ C5 Flow Cytometer; Biological safety cabinet; Electric pipette; Electric sterilizer; CO_2 incubator (37 ℃); Water bath; Inverted microscope; Cell counter

Activity protocol

(1) After taking photos of bright field and fluorescence images for each dish with

transf-ected HEK293T^{EGFP} cells, discard the cult-ure medium in the dishes and wash cells with 2 mL 1×PBS one time to remove remaining culture medium.

(2) Add 0.5 mL 0.25% trypsin-EDTA in the 60 mm culture dish and incubate at 37 ℃ for 3 min. Then disperse the detached cells gently several times with a transfer pipet and add 0.5 mL FBS to the dish to inactivate trypsin-EDTA. Add 4 mL DMEM to the same dish.

(3) Transfer the cell suspension to a 15 mL centrifuge tube and collect cells by centrifuging at room temperature at 200×g for 5 min. Discard the supernatant and resuspend the cells with 2 mL 1×PBS; gently disperse the cells with transfer pipet 10-20 times.

(4) Count the cell number and determine viability with 0.4% trypan blue staining and a hemocytometer or cell counter.

(5) Centrifuge the cell suspension at room temperature at 900 rpm for 5 min. Discard the supernatant and resuspend the cells in appropriate volume of 1×PBS to a density of 0.5×10^6 cells/mL.

(6) Transfer 1 mL of cell suspension to a 5 mL polystyrene round-bottom tube with cell-strainer cap and detect the transfection efficiency of cells on the BD Accuri^{TM} C5 Flow Cytometer.

Postlab Focus Questions

1. What is the major difference between transient transfection and stable transfection?

2. How do liposomes mediate the transfection of plasmid DNA into the cells?

3. What are the main applications of flow cytometry in life sciences?

(Written by Sun Caiyun)

Chapter 9 Cell Contamination

Cell culture is a common technique in life sciences and the quality of cells directly affects the credibility and reputability of scientific research. However, a significant proportion of cells are contaminated by various microorganisms, and some cell lines have even been mislabeled or replaced by cells from different individuals, tissues, or species, creating a problem that science has been unable to fully address. Papers published with misidentified or contaminated cells are misleading and bad for scientific innovation.

Section 1 Types of Cell Contamination

1. Mycoplasma

Mycoplasma, a bacterium without a cell wall structure that is only 50-300 nm in diameter, is currently considered the smallest prokaryotic microbe. Mycoplasma exists in a wide range of environments and may be found on the skin surface, nasal cavity, mouth, and various other parts of the experimenter. Because of its small size and absence of cell walls, it is able to pass through the filter membrane during media or reagent filtration processes and is resistant to antibiotics commonly used against cell walls. Mycoplasma contamination is not easily detected under an optical microscope, and, unlike bacterial contamination which causes rapid cell death, the medium does not change significantly, so it is often overlooked by experimenters. However, as culture time is prolonged, cells contaminated by mycoplasma deteriorate and proliferate slowly. Due to their parasitic or saprophytic life, the cells' antigenicity, metabolic behavior, and genetic material will change over time if they are not cleared of infection. At the same time, because of the poor state of the cells, results of subsequent cell experiments cannot be repeated. Therefore, researchers should regularly check cells for mycoplasma contamination.

2. Bacterial contamination

The most obvious feature of a culture contaminated with bacteria is becoming orange and cloudy because of the proliferation of bacteria, which can be seen suspended like fine sand by gently shaking the dish under the microscope, while cells essentially die and float in the medium. Depending on the type of bacterial contamination, different bacterial shapes can be observed under a high-power microscope, such as bacillus and streptococcus.

First, the rapid bacterial proliferation leads to cell death. Second, it is difficult to remove bacteria, even with double antibodies. Therefore, it is recommended to dispose of cells contaminated by bacteria as soon as possible to prevent contamination expansion and to revive new uncontaminated cells after screening with reagents and instruments (Fig. 9-1).

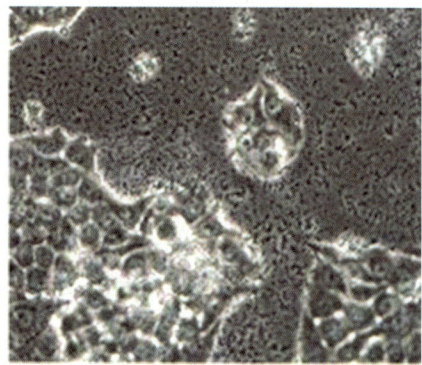

Fig. 9-1 Bacterial contamination in cell culture

3. Fungi pollution

Fungal contamination is mostly caused by mold contamination. Cell culture medium contaminated by mold remain clear and transparent, unlike the culture medium polluted by bacteria, but can still be distinguished. If the mold proliferation rate is fast, the naked eye can see a cluster of hyphae floating in the medium. The mycelium of mold can be observed clearly under the microscope, and although there are cells, the cells basically stop growing and are in an extremely poor state. In the case of fungal contamination with a slower growth rate, the mycelium also grows slowly over time and long and thin black filaments can be observed under the microscope (Fig. 9-2).

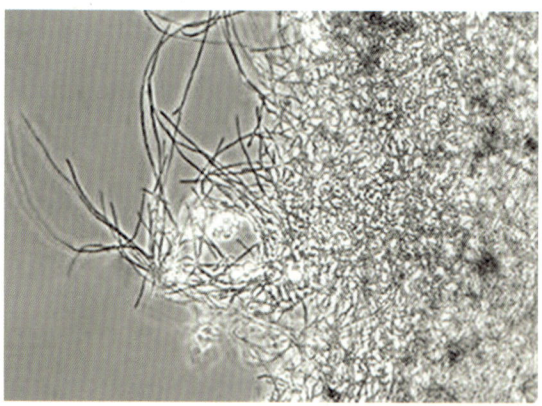

Fig. 9-2 Fungal contamination in cell culture

Mold contamination is most likely due to being left in the incubator for a long period of time where the appropriate environment led to mold growth and transfer to the cell medium. It is very difficult to save the cells after they have been contaminated by mold. It is recommended to discard mold contaminated cells immediately and carefully clean the incubator and screen operating instruments for contamination. If the cell line is more precious, amphotericin or mycetin can be used to treat the cells, but the effect is general and the antibiotics are generally used for prevention. Since ultraviolet irradiation cannot kill fungi, 75% ethanol and copper sulfate solution can be used regularly to wipe the inner wall of the incubator and each layer of the barrier: if the situation is more serious, formaldehyde fumigation can be used. Also, because fungi are not heat-resistant, a cell incubator that can be heated at high temperature could be purchased and subjected to high temperature heating to remove mold that regularly remains inside the incubator.

4. Cell cross-contamination

Cell cross-contamination (CCC) is a phenomenon in which cells of one culture are mixed into another cell culture, sometimes leading to the replacement of the original cells during the continuous culture process, particularly when the contaminant grows faster than the original line. According to incomplete statistics, over 15% of human cell lines are affected by CCC. Worryingly, many researchers do not know there is a problem with the cell lines they are using, which is bad for scientific development. At present, cells with CCC are still spreading. HeLa cells are perhaps the most famous example of a CCC cell line overtaking and then masquerading the original.

In the 1950s and 1960s, many continuous cell lines were unknowingly cross-contaminated with other cell lines, including HeLa cells. In the 1970s and 1980s, as many as one in three cell lines deposited in cell repositories were imposters. This cross-contamination was only uncovered with the development of suitable genetic markers beginning in 1967. Indeed, several "unique" cell lines in the ATCC turned out to be HeLa cells upon further study. Despite the confirmation of their HeLa cell origin, cytogenetic analysis suggested that there were differences among these HeLa-derived cell lines. Several of these cell lines possess unique properties. However, these cell lines should not be used as functional models of their claimed tissues of origin. Therefore, it is necessary to ensure the accuracy of cells before carrying out a project. In recent years, several research articles on CCC have been published in prominent journals. Since 2013, a series of prominent academic journals, represented by Nature, decided to require cell line authentication before submitting research papers[1].

A lack of awareness of standard operating procedures and poor cell culture technique are the main reasons behind CCC. There are four ways to cause CCC. First, two or more kinds of cells are maintained simultaneously. Second, different cell lines share the same cell culture medium and reagents. Third, during cell culture, cells are disseminated by operating devices such as pipettes. Fourth, cell contamination through the water bath.

CCC not only invalidates the experimental results, but also misleads the research direction. Therefore, learning excellent cell culture technique ensures high quality cells, thus ensuring the integrity of scientific research.

Section 2　Detection of Contamination

Part 1. Test methods for bacteria, fungi, and yeast

Most bacterial contamination occurs within a few days and is typically obvious to the naked eye. Distinct changes to the medium such as turbidity, presence of particles visible in suspension, and a rapid decline in pH (yellow color, indicating acidity) are all indicators of bacterial contamination. Fastidious bacterial species that grow very slowly can be difficult to detect.

Fungal contaminants may or may not cause a change in the pH of the medium and can be distinguished from bacteria by checking for the presence of filamentous structures in the suspension. Yeast cells are larger than bacteria, but may not appreciably change the pH of the medium, and will appear as separate round or ovoid particles. Microbacterial media, which can be used to test for bacterial and fungal contamination, include blood agar, thioglycollate broth, tryptic soy broth, BHI broth, Sabouraud broth, YM broth, and nutrient broth with 2% yeast extract.

However, some microbial contamination is not apparent. For example, the use of antibiotics can suppress bacterial growth and mask contamination. Some viral infections do not alter the morphology of the cells, and detection of mycoplasma contamination requires specific assays.

Part 2. Test methods for mycoplasma contamination

(1) DNA staining. DNA dyes (such as Hoechst 33258 and DAPI) are used to directly stain the cells (Fig. 9-3). Due to the presence of nucleic acids in mycoplasma, fluorescent dyes can also bind mycoplasma during the staining. Normal cells show fuller and brighter fluorescence with oval nuclei, while cells with mycoplasma contamination showed scattered and fragmented fluorescence. The results of DNA staining is the most direct and intuitive but there are some disadvantages. When the degree of mycoplasma contamination is low, the staining result is difficult to judge.

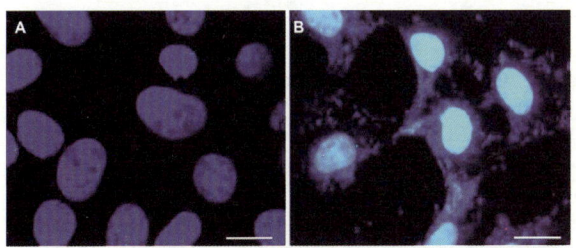

Fig. 9-3　Hoechst 33258 staining for mycoplasm
　　A. staining of healthy cells
　　B. staining of mycoplasma contaminated cells. Scale bar: 20 μm

(2) PCR detection. PCR detection is commonly used in various laboratories. Its principle is to design primers according to the conserved region of the mycoplasma 16S rRNA gene and amplify this genomic fragment of mycoplasma by PCR. The PCR product can be identified by agarose gel electrophoresis or the color change of reactant to the naked eye. PCR detection is convenient, fast, and sensitive; but the procedure needs to be relatively sterile to avoid false positive results caused by contamination during the detection process[2].

Part 3. Test methods for CCC

More recently, ATCC and other cell repositories have used DNA polymorphisms, enzyme polymorphisms, HLA typing, and karyotyping to confirm the identity of cell lines. One of the most reliable methods to study DNA polymorphisms is by profiling short tandem repeats (STR, Fig. 9-4). STR are usually a segment of DNA repeats composed of 3-7 base units in the genome. Because the number of core unit repeats is highly variable and abundant among individuals, it is a favorable tool for personal identification. Base on the polymorphism of allele fragment length in different cells, it can be determined if there is cross-contamination between cells[3].

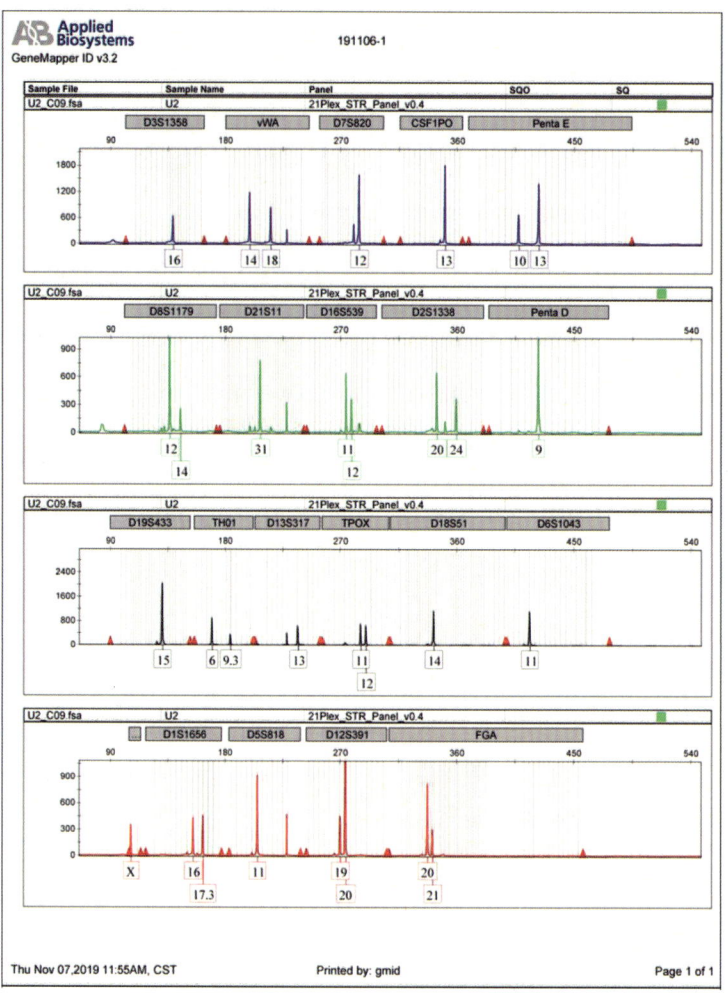

Fig. 9-4 STR profiles of osteosarcoma U2-OS cells. Amplicons were generated using the Applied Biosystems platform and separated by capillary electrophoresis

Section 3 Removal of Contamination

Part 1. Microbial contamination

Eliminating contamination from a cell line is time consuming and does not always work. Discarding the culture and starting over is preferred. However, if the cells are unique and irreplaceable, one should first identify the contaminant and select a suitable antibiotics for treatment. It is best to test the contaminating microbe for antibiotic sensitivity prior to treatment; this allows for a shorter treatment time and limits exposure of the cell line to potentially damaging reagents.

In the case of mycoplasma contamination, the state of cells becomes worse, so if the test result of mycoplasma is positive, mycoplasma should be removed immediately to prevent more cells from being contaminated. Because cells maintained in the laboratory are often cultured in the media containing antibiotics, such as penicillin and streptomycin, the contamination might be antibiotic-resistant. Therefore, commercial reagents specific for mycoplasma removal are recommended, such as tetracycline derivative, large ring lactone class, and fluoroquinolone antibiotics.

The "diseased" cells should be cultured for 1 to 2 weeks in the presence of the appropriate antibiotics, and then cultured without antibiotics for 1 to 2 weeks. At this point, the cell line should be retested with a very sensitive method to ensure that the culture is clean. Periodic retesting should be employed to make sure that the contaminant does not reappear.

However, because antibiotics may be toxic to cells, treatment may lead to a selected subpopulation that no longer exhibits characteristics of the parental cells. Therefore, it should be considered to discard the contaminated cells and replace them with cells from different uncontaminated batches for experiments to ensure cells are in the best state.

Part 2. CCC

If CCC occurred, generally, the cells should not be used. If the cells are unique and precious, cell separation using specific antibodies can be considered, but the efficacy is limited.

Appeal

Contamination of cells in culture can arise from many sources including reagents, supplies such as pipettes and culture vessels, equipment such as tissue culture hoods, waterbaths and incubators, and laboratory personnel. While the potential for contamination is constant, the risk can be reduced or eliminated with proper precautions:

(1) Do not enter the cell culture room when tired or unable to concentrate.

(2) Clean the superclean bench and equipment with disinfectant thoroughly before and after using the culture media and treating the cells.

(3) Use only reagents of known quality and sterility. Before doing cell experiments, plan carefully to make sure you have enough reagents and supplies. Do not share reagents with others. Reagents used for each cell line should be specialized, if possible.

(4) Purchase cell lines from a certified, qualified cell bank. Do not transfer cells between research teams without rigorous testing. Quarantine new cell lines until they are tested to be free from contamination.

(5) Examine cell status by microscope before beginning experiments.

(6) Operate only one cell line or one lineage of cells at the same time. Do not operate two or more kinds of cells at the same time.

(7) Use a pipette only once. Do not place used pipettes back into reagent bottles.

(8) Authenticate cell lines every six months if possible.

(9) Properly train cell culture personnel.

Section 4 Know More

Small body, great contribution

Henrietta Lacks, a small black American woman (Fig. 9-5), has been dead for almost 70 years, but her cells are still alive. HeLa cells were derived from Lacks' cervix and were named as such from the first two letters of her first and last names.

Fig. 9-5 Henrietta Lacks (1920-1951)

In January 1951, Lacks, at the age of 30, found a lump in her abdomen and exhibited vaginal bleeding. She went to John Hopkins hospital and was diagnosed with cervical cancer. Without her or her family's knowledge, her doctor collected cancer tissue from Lacks and cultured the tissue in a petri-dish that day. In October 1951, Lacks died, but the cells the doctor collected from her did not die. Instead, the cells showed active growth and became the first immortalized human cell line. Her cells have been living in a different form in laboratories around the world ever since.

Scientists found HPV, the human papilloma virus that causes cervical cancer, in HeLa cells and developed a vaccine against it. HeLa cells have also helped scientists achieve genomic research, cell cloning, lay the foundation for animal cloning, gene therapy, *in vitro* fertilization, and stem cell research and applications. In addition, people also use HeLa cells to study oncogenes and tumor suppressors, develop drugs to treat herpes, leukemia, influenza, hemophilia, and Parkinson's disease, and discover and use green fluorescent protein to explain the secret of long-life. More than 145,000 papers have been published using HeLa cells according to PubMed, a medical and biological database. Since the beginning of the 21st century, five research achievements based on HeLa cells have won the Nobel Prize, including "the discovery of HPV" and "the discovery and development of green fluorescent proteins."

So far, HeLa cells have lived in the single-cell state for more than 18,000 generations. The wide application of HeLa cells has made great contributions to human beings, not only in research, but also in bioethics. As explorers of life sciences, we should be in awe of life.

References

[1] HORBACH S, HALFFMAN W, et al. The ghosts of HeLa: How cell line misidentification contaminates the scientific literature [J]. PLoS One, 2017, 12: e0186281. doi: 10.1371/jour-nal.pone.0186281.

[2] CORRAL V C, AGUILAR Q R, et al. Cell lines authentication and mycoplasma detection as minimun quality control of cell lines in biobanking [J]. Cell tissue bank, 2017, 18: 271-280.

[3] BIAN X, YANG Z A, et al. Combination of species identification and STR profiling identifies cross-contaminated cells from 482 human tumor cell lines [J]. Sci Rep., 2017, 7: 9774. doi: 10.1038/s41598-017-09660-w.

Postlab Focus Questions

1. Why is it important for researchers to thoroughly characterize their cell lines?

2. How would you describe "faulty cell culture techniques" that can often lead to cross-contaminated cell lines?

(Written by Zhang Yan)

Chapter 10　Genome Editing by CRISPR/Cas9

Background Reading

Disrupting a gene to evaluate its effect on an organism's phenotypes is an indispensable tool in molecular biology. Such techniques are critical for understanding how a gene contributes to the development of organisms. In the past decade, researchers have hypothesized that by exploiting the endogenous cellular DNA repair pathways, one could create precise edits at a desired locus in the genome, termed genome editing. Double-strand breaks (DSBs) are toxic to cells, thus organisms have evolved mechanisms to repair these lesions. Scientists proposed that by generating a targeted DSB at a site of interest, errors may occur during the repair to result in a mutation at that site. Additionally, endogenous double-strand break repair pathways could also stimulate incorporation of exogenous DNA, creating specific researcher-designed edits. Thus, researchers started to identify ways to target enzymes that generate double-strand breaks, called nucleases, to specific regions of the genome. An RNA-directed nuclease from a bacterial immune system called Cas9 has proven to be an easily programmed enzyme that can create double-strand breaks in eukaryotes[1-2].

Section 1　CRISPR/Cas9 System

Background Reading

Prokaryotes have defense mechanisms against viral and plasmid cellular invaders, just like multicellular organisms. One of these defense mechanisms is an adaptive immune system found in many bacteria and most archaea called Clustered Regulatory Interspaced Short Palindromic Repeats (CRISPR), along with the CRISPR-associated proteins known as Cas proteins. By integrating DNA sequences identical to those of past invaders into their gen-ome, bacteria and archaea are able to generate a cellular memory of past invaders. These acquired sequences allow the bacteria or archaea to recognize viral or plasmid invaders as non-self and degrade the invading sequence, essentially functioning as an adaptive immune system for prokaryotes. CRISPR immunity is characterized by distinct phases. First, during the adaptation phase, bacteria or archaea gain a cellular memory of the invading virus or plasmid. Short sequences of the viral or plasmid genomes are integrated into the CRISPR locus of the bacterial or archaeal genome (Fig. 10-1A). These CRISPR loci were first identified by scientists working in the fermentation industry, where prokaryotes are essential for the production of fermented products. Through comparative genomic analysis of different S. thermophiles strains (a microbe used in producing yogurt), scientists identified a highly variable locus in the genome of these bacteria. This highly variable region had two distinct features: many non-contiguous repeats and the variable sequences termed spacers that separated them. Upon closer inspection, researchers found that the spacer sequences matched those found in phage (viruses that infect bacteria) genomes. Interestingly, when researchers compared phage resistant and phage sensitive, S. thermophilus, the phage resistant bacteria, had spacer sequences that matched regions of that phage's genome. Thus, spacer content correlated with phage resistance, leading to the model that short regions of the invader's genome are integrated into the CRISPR loci as a spacer, separated by repeat sequences, resulting in a cellular memory of previous infections (Fig. 10-1 A). After the acquisition of spacers, RNA, termed the CRISPR RNA (crRNA), is generated from spacers at the CRISPR locus and loaded onto a Cas protein. crRNA directs the Cas protein to recognize invading sequences and cleave the incoming phage or plasmid DNA (Fig. 10-1B). Three different types of CRISPR-Cas systems have been identified in bacteria and archaea: Type I, Type II, and Type III. Each system utilizes a different mechanism to generate crRNA and Cas proteins that catalyze the nucleic acid cleavage. Here, we will focus on the Type II CRISPR system, which has been the most commonly adapted system for genome editing due to its simplicity in requiring only one Cas protein, Cas9, and two RNA components. To generate the crRNA, the CRISPR locus is transcribed to generate a long RNA molecule with sequences homologous to those of past invaders. This RNA molecule is termed the pre-crRNA. A second RNA from a genomic locus upstream of the CRISPR locus is also transcribed. This RNA is called the trans-activating CRISPR RNA (tracrRNA). The tracrRNA has a region that is complementary to the repeat region of the CRISPR locus, and binds the newly transcribed pre-crRNA to create a double-stranded RNA which gets cleaved by RNaseIII (an enzyme that recognizes and cuts double-stranded RNA) to produce a crRNA-tracrRNA; crRNA-tracrRNA complex containing just one spacer sequence (Fig. 10-1A). This RNA complex then associates with a single Cas9 protein to create an active ribonucleoprotein (RNP) complex (Fig. 10-1A). Once the crRNA-tracrRNA is bound by Cas9, Cas9 is activated and can cleave invading nucleic acid sequences (interference, Fig. 10-1B).

Cas9 is termed an RNA-guided endonuclease because it cleaves DNA at sequences that bind to the crRNA of the Cas9 RNP. Searching the invading DNA for sequences complementary to the crRNA occurs through Cas9 binding to sequences in the invading viral or plasmid genome termed Proto-spacer Adjacent Motifs or PAMs. Different Cas9 proteins from different species of bacteria or archaea recognize different PAM sites. To date, S. pyogenes Cas9 (SpCas9), which recognizes a 5′-NGG-3′ PAM, is the most commonly used for genome editing (Fig. 10-2A). Two critical arginine residues in SpCas9, Arg1333 and Arg1335, interact with the guanine nucleobases of the PAM on the noncomplementary strand (Fig. 10-2B). This interaction between the guanines of the PAM and the arginines of SpCas9 positions the phosphate of the DNA backbone 5′ to the PAM to interact with a phosphate-lock loop in Cas9 to facilitate DNA strand unwinding. If the DNA is complementary to the guide

RNA, an RNA-DNA hybrid forms, called an R loop, and cleavage follows. DNA cleavage is mediated by two different Cas9 nuclease domains: the HNH domain nicks the DNA strand that is complementary to the crRNA and the RuvC-like domain nicks the strand that is not complementary to the crRNA (Fig. 10-2C). Cas9 cleaves the DNA three base pairs upstream of the PAM, resulting in a blunt-end cleavage of DNA (Fig. 10-2D). Cleaving the DNA is deleterious to the invading plasmid or virus, resulting in their degradation and protecting the host against these invaders.

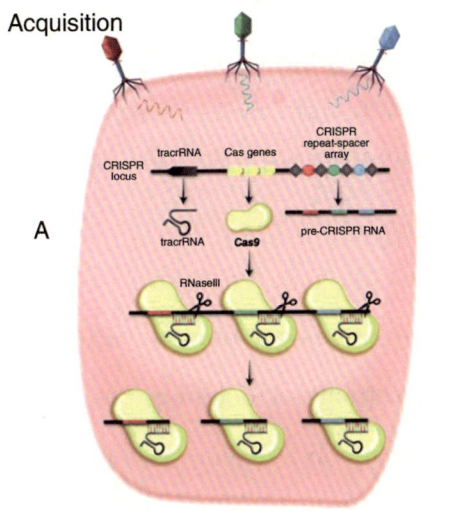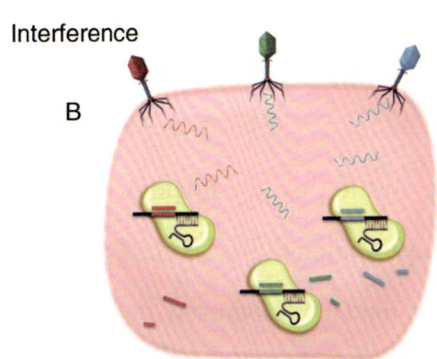

Fig. 10-1 CRISPR/Cas9 mediated acquired immunity in prokaryotes

During the acquisition phase (A), cellular inv-aders, such as a phage virus, inject nucleic acid sequences into the host cell. After infection, novel DNA seq-uences from the cellular invaders are incorporated into the host CRIPSPR locus as spacers (colored circles) flanked by repeat sequences (gray diamonds). As a result, when the CRISPR locus is transcribed, the pre-CRISPR RNAs (crRNAs) encode the newly acquired protospacer sequences. The pre-crRNA is cleaved to produce individual crRNAs that will associate with Cas proteins. The Cas protein utilizes the crRNAs as guides to silence foreign DNA that matches the crRNA sequence (B, interference phase). As a result, the second time a bacteria encounters the same foreign DNA, the crRNA/Cas9 complex is able to identify and degrade the foreign DNA.

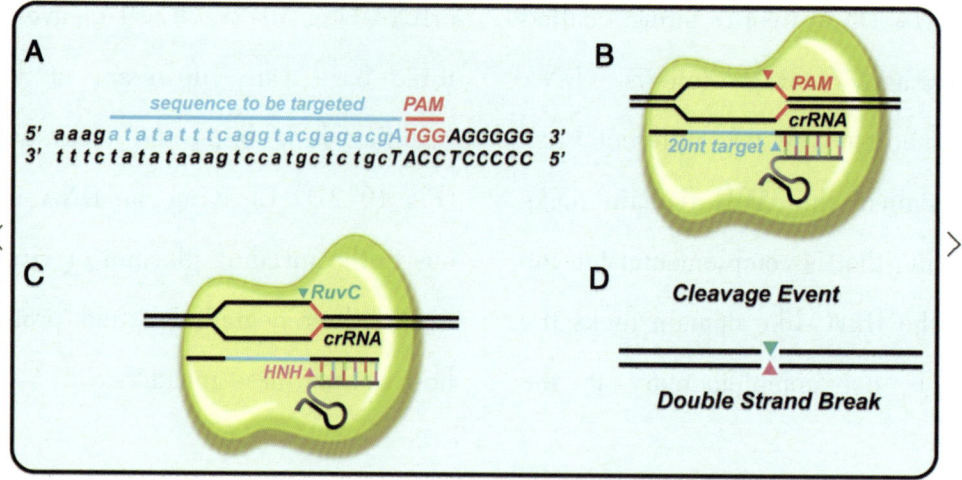

Fig. 10-2　Mechanisms of CRISPR/Cas9 mediated genome editing

(A) Sequence of a targeted genomic locus in relation to the PAM (5′-NGG-3′) site. (B) Cartoon representation of crRNA, tracrRNA, and Cas9 protein ass-embly. (C) Cas9 contains two nuclease domains, RuvC and HNH, which each cut a different strand of the DNA, resulting in a blunt-end cleavage (D).

After initial characterization of the CRISPR/Cas9 microbial immune system, molecular biologists recognized how it could be exploited for precise genome editing in eukaryotes. In response to Cas9 induced double-strand breaks, cells employ one of two DNA repair pathways to repair the damage: either non-homologous end joining (NHEJ) or homology-directed repair (HDR) (Fig. 10-3). NHEJ can occur through canonical NHEJ (C-NHEJ), which ligates or essentially "glues" the broken ends back together. Additionally, there is an alternative end joining pathway (alt-NHEJ) in which one strand of the DNA on either side of the break is resected to repair the lesion. Both of these repair methods are error-prone, meaning that the lesion is repaired imperfectly, resulting in insertions or deletions (Indels, Fig. 10-3). Alternatively, if there is a nearby DNA molecule with homology to the region around the double-strand break is nearby, then the homologous DNA can be used as a template to repair the break through the homology-directed repair (HDR) pathway. This form of repair can be exploit-ed to introduce precise edits or large insertions or deletions by introducing a donor template for repair (Fig. 10-3). Thus, by cutting at a specific locus and taking advantage of the cellular DNA repair pathways provides the potential to generate targeted mutations and insert sequences of interest.

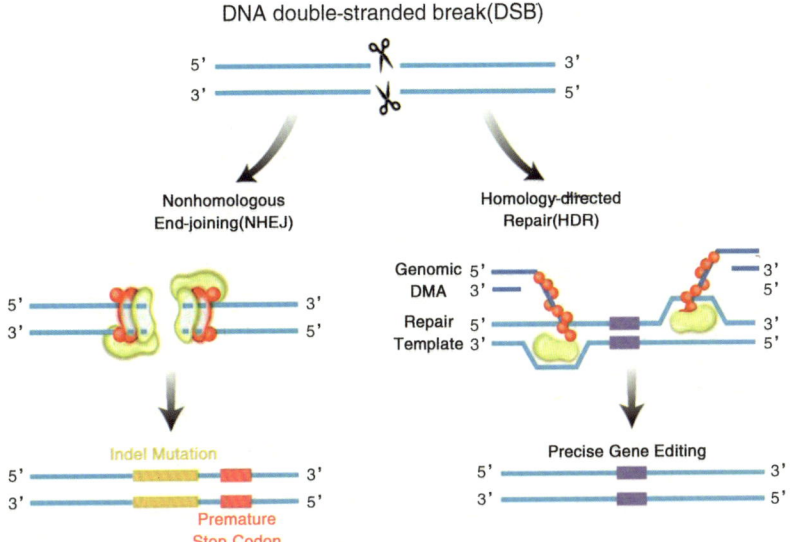

Fig. 10-3 Cas9 induced double-strand breaks can be repaired by either nonhomologous end-joining (NHEJ) or homology-directed repair (HDR)

NHEJ results in random insertions, deletions, and indels. HDR results in prec-ise researcher-designed edits. To achieve HDR, the researcher also introduces a repair template containing the desired edit to be used by HDR repair machinery of the cell to repair the induced double-strand break.

To adapt CRISPR for genome editing in eukaryotes, researchers first characterized Cas9 and the role of the crRNA-tracrRNA complex. Through *in vitro* studies utilizing purified Cas9 to cut a DNA template either with or without the tracrRNA, researchers found that the tracrRNA is required for DNA cleavage by Cas9. Additionally, researchers found that the crRNA and tracrRNA could be combined into a single guide RNA or sgRNA, limiting the number of components that must be introduced into the cell (Fig. 10-4A). Next, three different studies showed that SpCas9 expression with a sgRNA precisely targets Cas9, resulting in a cut at a researcher-specified location in the mouse or human genome and demonstrating the feasibility of CRISPR/Cas9 as a eukaryotic genome editing tool. One major advantage of the CRISPR/Cas9 system, as compared to conventional gene targeting and other programmable endonucleases, is the ease of multiplexing, by which multiple genes can be mutated simultaneously by simply using multiple sgRNAs targeting different genes (Fig. 10-4B). In addition, when two sgRNAs that flank a genomic region are used, the intervening region can be deleted or inverted (Fig. 10-4B). For small modifications, such as incorporation of point mutations, defined indel mutations, insertion of a short sequence such as a loxP site or an epitope tag, single-stranded oligodeoxy-nucleotides (ssODNs) can be used as donor DNA by the HDR pathway (Fig. 10-4B). In this design, donor ssODNs are designed to carry homologous sequences flanking the mutation site and their total size can be up to 200 nt. When DNA of larger sizes is to be introduced into a target site, a double-strand-ed donor plasmid carrying the transgene flanked by homology arms is used (Fig. 10-4B).

Because of the ease of use, the CRISPR/Cas9 system has swiftly become the most commonly used tool for efficient genome editing in bacteria, plants, cell lines, primary cells, and tissues. Impressively, direct introduction of CRISPR/Cas9 into the zygote leads to efficient genetic modification of the genome in early embryos, which, when brought to term, develop into genetically modified animals[1-2].

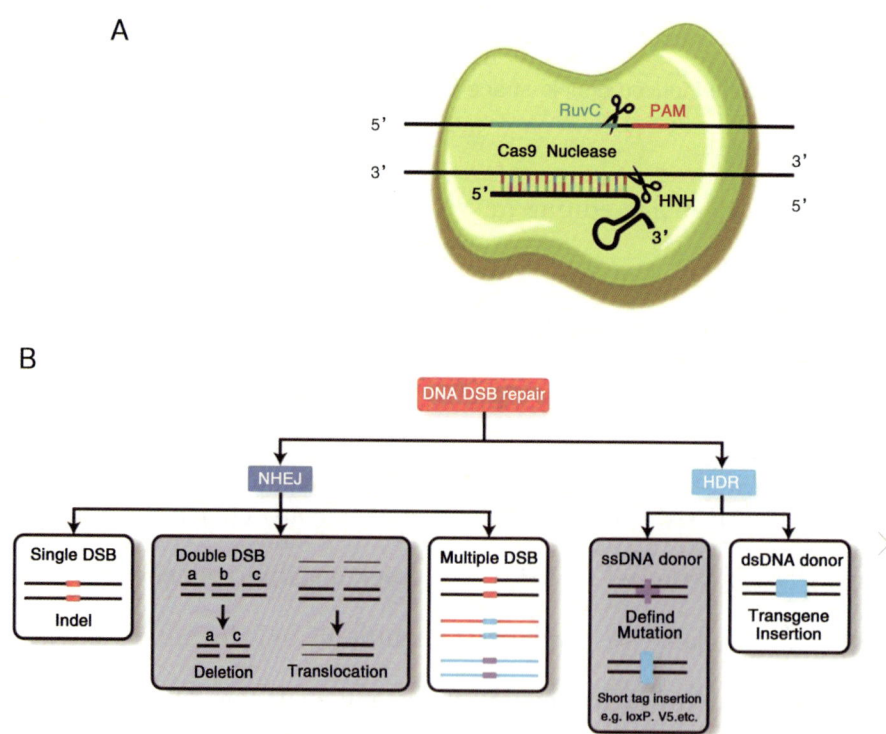

Fig. 10-4　CRISPR/Cas9-mediated genome editing

(A) The structure of the Cas9/sgRNA complex binding to target DNA. Cas9 binds to specific DNA sequences via base-pairing between the guide sequence on the sgRNA (pink) with the DNA target (gray). The protospacer adjacent motif (PAM) is downstream of the target seq-uence. (B) CRISPR/Cas9-mediated double-stranded DNA breaks are repaired by endogenous DNA repair machinery: non-homologous end joining (NHEJ) or homology-directed repair (HDR). Various genetic modificat-ions can be generated through these two pathways.

Targeted deletion of the integrated EGFP gene in HEK293^{EGFP} cells with CRISPR/Cas9

As previously mentioned, when two sgRNAs flanking genomic region are used, the intervening region can be deleted or inverted (Fig. 10-4B). Here, we practice how to knockout the integrated EGFP gene in HEK293^{EGFP} cells with a paired Cas9/sgRNA system (Fig. 10-5).

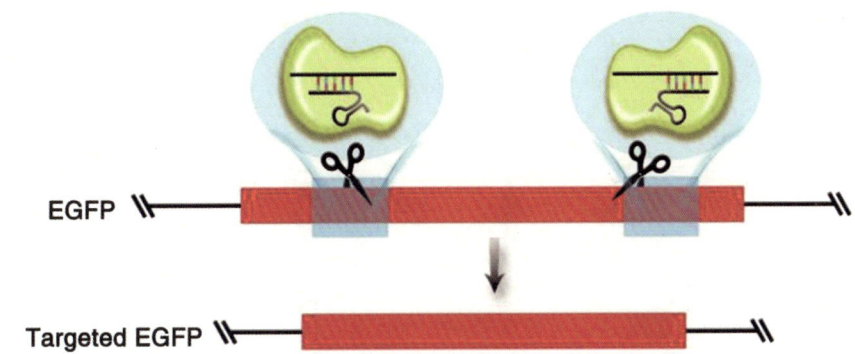

Fig. 10-5 Schematic diagram of the targeted deletion of EGFP gene by CRISPR/Cas9

Diverse RNA-programmable CRISPR/Cas enzymes and extended applications

At the tail end of 2015, type II CRISPR/Cas systems expanded to include a number of candidate systems, which were later designated as type V CRISPR/Cas12a (formerly Cpf1) and type VI CRISPR/Cas13a (formerly C2c2, Fig. 10-6). Today, SpCas9 shares the spotlight with a variety of Cas9 homologs, DNA targeting Cas12, and RNA-targeting Cas13, all of which are programmable RNA-guided nucleases. This inherent programmability present in a variety of naturally evolved systems extends the applications of CRISPR/Cas beyond precision genome editing.

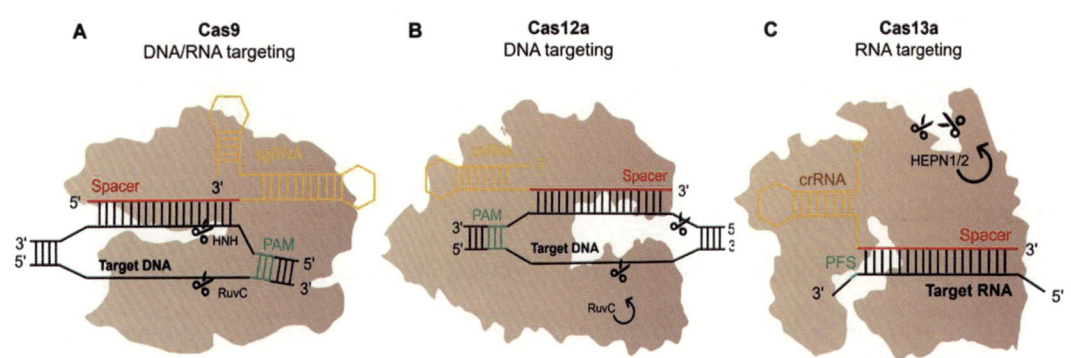

Fig. 10-6 Schematic of type II CRISPR-Cas systems

Cas9 and Cas12a are used to induce dsDNA breaks for genome editing. NCas9 can be fused to base editors to modify nucleotides in dsDNA for genome editing without introducing a dsDNA break. DCas9 can be fused to transcriptional activators, repressors, or epigenetic modifiers to regulate transcription. Cas9 and Cas13a can be used for targeted RNA interference. Cas13a fused to base editors can modify nucleotides in RNA. DCas9 or dCas13a can be fused to GFP to visualize DNA or RNA (Fig. 10-7).

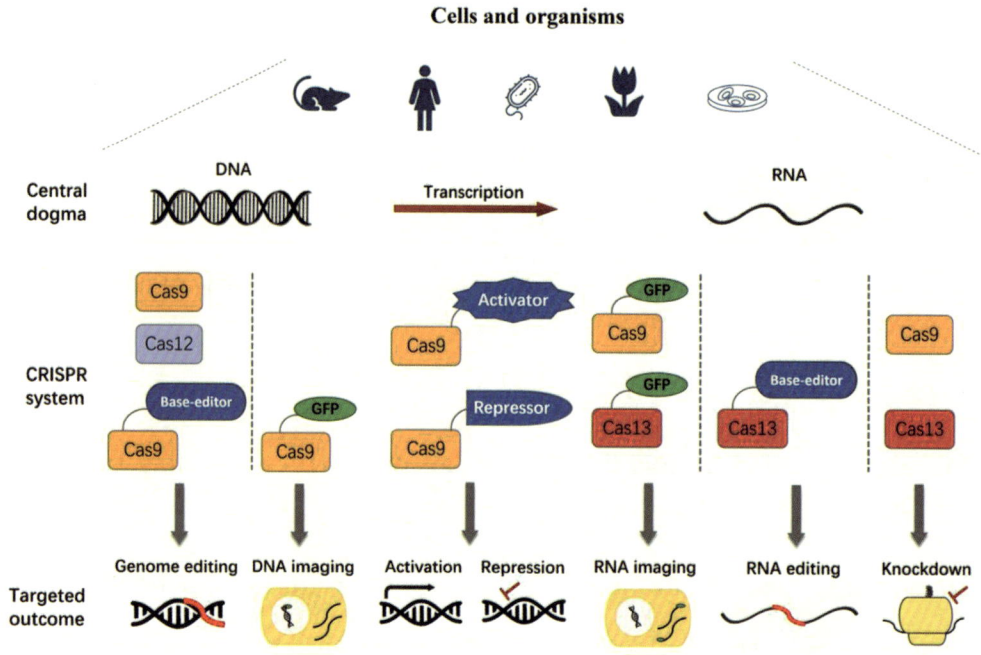

Fig. 10-7　CRISPR/Cas systems allow genetic manipulations across the central dogma

Pathogen detection by CRISPR/Cas systems

Surprisingly, along with their divergent sequences, Cas12a and Cas13a also deviate in their catalytic activities in comparison to Cas9. Upon binding to a target RNA, Cas13a transforms into a nonspecific endoribonuclease that can degrade single-stranded RNA sequences supplied either in cis or in trans, with the target. This so-called collateral cleavage activity diverged from the known activity of the other type II Cas endonucleases, which were thought to cleave only at specific sites within the target. However, a new mechanistic study has shown that Cas12a target binding also triggers nonspecific collateral cleavage, this time against single-stranded DNA supplied in trans. This activity be exploited to amplify target detection through the collateral cleavage of a reporter nucleic acid. Owing to the long-term and multiple turnover nature of collateral cleavage, the signal could be amplified over time to ensure detection, even in the presence of a small amount of the target sequence, based on the isothermal amplification of input nucleic acid. The resulting DNA can be transcribed (txn) for Cas13-based detection, or detected directly by Cas12a. Reporter ssRNA or ssDNA are cleaved by Cas13 or Cas12a, respectively, producing a fluorescent signal (Fig. 10-8)[3].

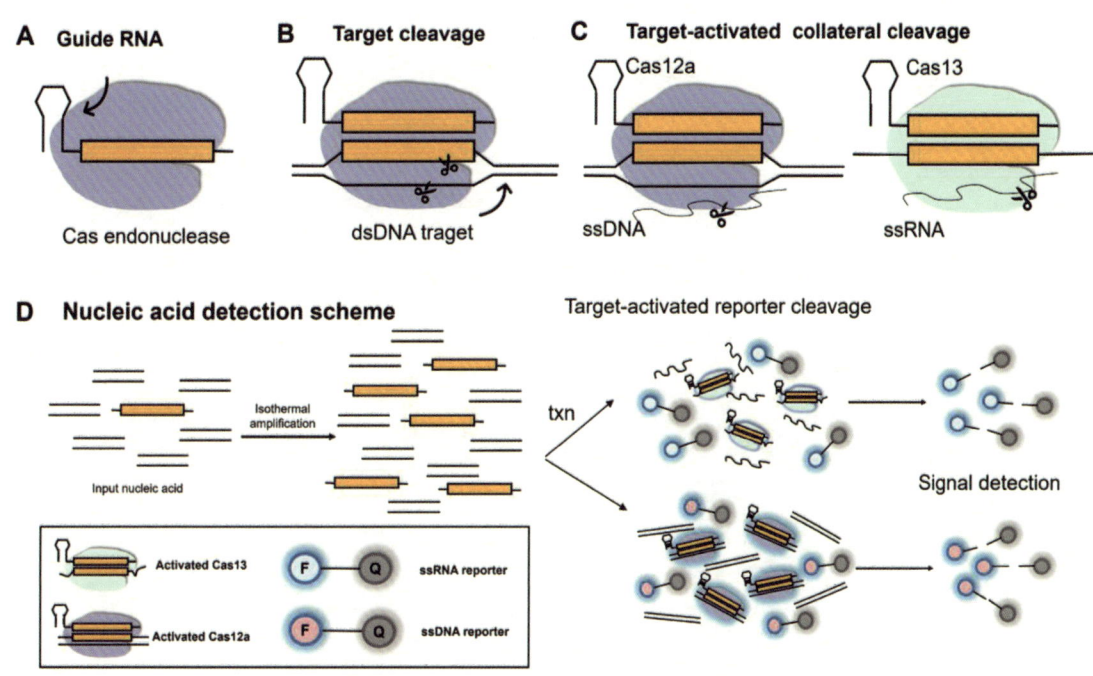

Fig. 10-8 Overview of Cas endonuclease activity and nucleic acid detection systems

Cas13a-based RNA detection opens the possibility of diagnostic applications by detecting nucleic acids associated with pathogens or diseases (see Know More).

Section 2 Cloning a gRNA into a Vector

Objectives

❖ Definition of the different components necessary for genome editing by CRISPR/Cas9

❖ Construction of two Cas9/sgRNA expression vectors

❖ Identification of the desired transformants by colony PCR

Materials

Vector: pX459

Competent cell: *E. coli* JM109

Reagents: *Bbs* I; T4 PNK; T4 ligase; SOC medium; LB plate; 2 × Taq Mix; Primers; 50 × TAE; SYBR Green; Agarose; SYBR Green; DNA marker (100 bp); Loading buffer.

Supplies: Pipette tips; Microcentrifuge tubes; Ice bucket; Marker pen; Floater.

Equipment: Pipette; Microcentrifuge; Thermal cycler (PCR machine); Shake incubator; Water bath; Gel electrophoresis apparatus.

Procesure

See Fig. 10-9.

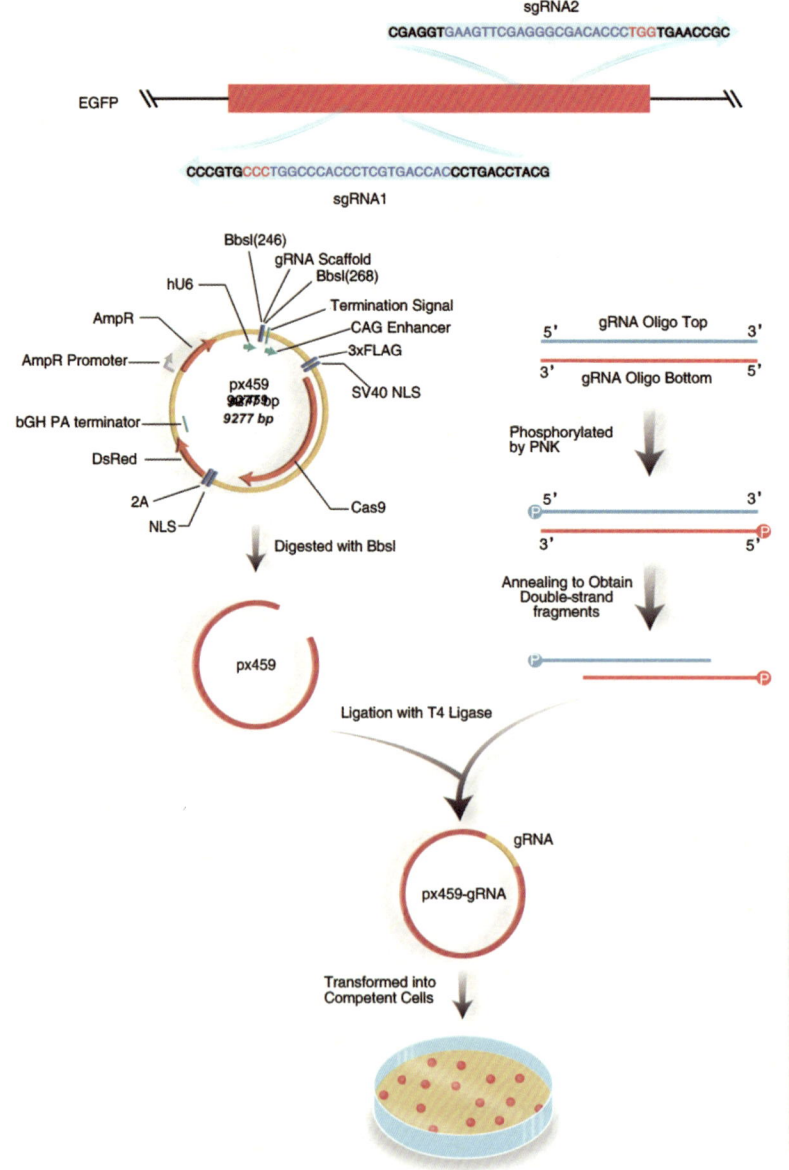

Fig. 10-9 Schematic of cloning a gRNA into a Cas9/sgRNA coexpression vector

Activity Protocol

(1) Resuspend the top and bottom strands of gRNA oligos to a final concentration of 100 μM. Prepare the following mixture (Table 10-1) for phosphorylating and annealing the gRNA oligos:

Table 10-1 Components required for phosphorylation of gRNA oligos

Component	Amount (μL)
gRNA oligo top (100 μM)	1
gRNA oligo bottom (100 μM)	1
T4 PNK	1
T4 Ligation buffer, 10 ×	1
ddH$_2$O	6
Total	10

(2) Phosphorylate and anneal the oligos in a thermocycler with the following parameters: 37 ℃ for 30 min; 95 ℃ for 5 min; ramp down to 25 ℃ at 5 ℃ min.

(3) Dilute phosphorylated and annealed oligos 1∶200 by adding 1 μL of oligo to 199 μL of room temperature ddH$_2$O.

(4) Cloning the gRNA oligos into pX459. Set up a ligation reaction for gRNA oligos as described below (Table 10-2):

Table 10-2 Components required for cloning gRNA into pX459 vector

Component	Amount (μL)
pX459, 100 ng/μL	1
Diluted oligo duplex from Step 3	2
Bbs I	1
T4 Ligase	0.5
T4 Ligation buffer, 10 ×	2
ddH$_2$O	13.5
Total	20

Note: *Set up a non-insert, pX459-only negative control for the ligation.*

(5) Mix the components by thoroughly pipetting (Table 10-3) and incubate the ligation reaction in a thermocycler for 1 h using the following program:

Table 10-3 Program used for cloning gRNA into pX459 vector

Cycle number	Condition
1-6	37 ℃ for 5 min, 21 ℃ for 5 min

(6) Transform the ligation products into competent JM109 cells following the protocol from Chapter 4. Briefly, add 5 μL of the product from Step 5 to 50 μL of ice-cold chemically competent JM109 cells, incubate the mixture on ice for 10 min, heat-shock

the mixture at 42 ℃ for 90 s, and return it immediately to ice for 2 min. Add 500 μL of SOC medium, incubate at 37 ℃ and centrifuge at 220 rpm for 1 h, plate 200 μL of culture on an LB plate containing 100 μg/mL ampicillin. Incubate the plate overnight at 37 ℃. Transform 1 μL of pX459 into 20 μL of competent cells as a positive control, and transform 1 μL of pX459 digested with BbsI into 20 μL of competent cells as a negative control.

(7) Inspect the plates for colony growth. Typically, there are no colonies on the negative control plate (ligation reaction of BbsI-digested pX459 alone without annealed gRNA oligo insert), and there are tens to hundreds of colonies on the pX459-gRNA (gRNA inserted into pX459) cloning plates.

(8) From the pX459-gRNA cloning plate, pick three colonies and transfer each to a PCR tube containing PCR reaction components with a pipette tips. Pick one colony from the positive control plate as a control for the analysis.

(9) Set up the PCR reaction as follows (Table 10-4):

Table 10-4 Components required for colony PCR reaction

Component	Amount (μL)
Bacterial Colony	0
Forward Primer (100 nM)	1
gRNA oligo Bottom (100 nM)	1
Taq Mix, 2×	10
ddH$_2$O	8
Total	20

10. Run the PCR under the following program: denaturing at 95 ℃ for 5 min, 40 cycles of denaturing at 95 ℃ for 30 s, annealing at 60 ℃ for 30 s, and extension at 72 ℃ for 30 s, followed by an additional elongation at 72 ℃ for 10 min, and kept at 25 ℃ forever.

11. Run the PCR product on a 2% agarose gel.

Section 3　Purification of Cas9/sgRNA Coexpression Vectors

❖ Objective

♦ Purification of endo-free plasmid using a commercial kit

❖ Materials

Competent cell: *E. coli* JM109

Reagents: LB medium; E. Z. N. A.® Endo-Free Plasmid DNA Midi Kit; Isopropanol; Ethanol; 50 × TAE; SYBR Green; Agarose; SYBR Green; DNA marker (1000 bp); Loading buffer

Supplies: Pipette tips; Microcentrifuge tubes; Ice bucket; Marker pen

Equipment: Pipette; Shake incubator; Microcentrifuge (with cooling system); Water bath; Spectrometer; Gel electrophoresis apparatus

Activity Protocol

(1) Inoculate a 200 μL culture for each of pX459-gRNA1 and pX459-gRNA2 transformant into separate tubes of 50 mL LB medium containing 100 μg/mL ampicillin, respectively. Incubate the culture and shake it at 37 ℃ overnight. (culture of transformants was prepared in advance)

(2) Transfer 30 mL overnight culture to a 50 mL centrifuge tube.

(3) Centrifuge at 4,000 × g for 10 min at room temperature.

(4) Decant and discard the culture media.

Note: *To ensure that all traces of the medium are removed, use a clean paper towel to blot excess liquid from the wall of the tube.*

(5) Add 2.5 mL Solution I/RNase A. Vortex or pipet up and down to completely resuspend the cells.

Note: *RNase A must be added to Solution I before use.*

(6) Add 2.5 mL Solution II. Invert and rotate the tube gently 8-10 times to obtain a cleared lysate. This may require a 2-3 min incubation at room temperature with occasional mixing.

Note: *Avoid vigorous mixing as this will shear chromosomal DNA and lower plasmid purity. Do not allow the lysis reaction to proceed more than 5 min. Store Solution II tightly capped when not in use to avoid acidification from CO_2 in the air.*

(7) Add 1.25 mL cold N3 Buffer. Gently invert 10 times or until a flocculent white precipitate forms. This may require a 2 min incubation at room temperature with occasional mixing.

Note: *The solution must be mixed thoroughly. This is vital for obtaining good yields. If the mixture still appears viscous, brownish, or conglobated, more mixing is required to completely neutralize the solution.*

(8) Prepare a Lysate Clearance Filter Syringe by removing the plunger. Place the barrel in a tube rack to keep upright. Make sure the end cap is attached to the syringe tip.

(9) Immediately transfer the lysate from Step 7 into the barrel of the Lysate Clearance

Filter Syringe.

(10) Hold the Lysate Clearance Filter Syringe barrel over a clean 15 mL centrifuge tube and remove the end cap from the syringe tip.

(11) Gently insert the plunger into the barrel to expel the cleared lysate into the 15 mL centrifuge tube.

Note: *Some of the lysate may remain in the flocculent precipitate. DO NOT force this residual lysate through the filter.*

(12) Measure the volume of cleared lysate.

(13) Add 0.1 volume ETR Solution. Invert the tube gently 10 times.

(14) Incubate on ice for 10 min. Invert the tube several times during the incubation.

Note: *After addition of ETR Solution, the lysate should appear turbid, but it should become clear after incubation on ice.*

(15) Incubate the lysate at 42 ℃ for 5 min. The lysate should appear turbid again.

(16) Centrifuge at 4,000 × g for 5 min at 25 ℃. The ETR Solution will form blue layer at bottom of tube.

(17) Transfer the top aqueous phase (cleared lysate) to a new 15 mL or 50 mL tube, add 0.5 volume absolute ethanol (room temperature, 96%-100%). Gently invert 6-7 times. Incubate at room temperature for 1-2 min.

(18) Insert a HiBind® DNA Midi Column into a 15 mL collection tube.

Optional Protocol for Column Equilibration:

◆ Add 1 mL GPS Buffer to the HiBind® DNA Midi Column.

◆ Let sit at room temperature for 4 min.

◆ Centrifuge at 4,000 × g for 3 min.

◆ Discard the filtrate and reuse the collection tube.

(19) Transfer 3.5 mL cleared supernatant from Step 17 to the HiBind® DNA Midi Column.

(20) Centrifuge at 4,000 × g for 3 min.

(21) Discard the filtrate and re-use the collection tube.

(22) Repeat Steps 19-21 until all of the cleared supernatant has been transferred to the HiBind® DNA Midi Column.

(23) Add 3 mL HBC Buffer.

Note: *HBC Buffer must be diluted with isopropanol prior to use.*

(24) Centrifuge at 4,000 × g for 3 min.

(25) Discard the filtrate and reuse the collection tube.

(26) Add 3.5 mL DNA Wash Buffer.

Note: *DNA Wash Buffer must be diluted with 100% ethanol prior to use.*

(27) Centrifuge at 4,000 × g for 3 min.

(28) Discard the filtrate and reuse the collection tube.

(29) Repeat Steps 26-28 for a second DNA Wash Buffer wash step.

(30) Centrifuge the empty HiBind® DNA Midi Column at 4,000 × g for 10 min to dry the column matrix.

Note: *It is important to dry the HiBind® DNA Midi Column matrix before elution. Residual ethanol may interfere with downstream applications.*

(31) Transfer the HiBind® DNA Midi Column to a nuclease-free 15 mL centrifuge tube.

(32) Add 0.5-1 mL Endo-Free Elution Buffer directly to the center of the column matrix.

(33) Let it sit at room temperature for 3 min.

(34) Centrifuge at 4,000 × g for 5 min.

Note: *This represents approximately 65%-80% of bound DNA. An optional second elution will yield any residual DNA, though at a lower concentration. Alternatively, a second elution may be performed using the first eluate to maintain a high DNA concentration.*

(35) Quantify the concentration of plasmid on the BioTek microplate reader

(36) Run the purified plasmid DNA on a 1% agarose gel.

(37) Store DNA at -20℃.

Section 4
Cotransfection of Cas9/sgRNA Coexpression Vectors into the Cells

Objectives

❖ Transfection of plasmids into human cells using lipofectamine

❖ Determination of transfection efficiency by flow cytometry

Materials

Vector: pX459-gRNA1; pX459-gRNA2

Cell: HEK293^{EGFP}

Reagents: DMEM culture medium; PBS; Opti-MEM Medium; P3000™ Reagent; Lipofectamine™ 3000 Reagent; 0.25% trypsin-EDTA

Supplies: Pipette tips; Microcentrifuge tubes; 60 mm dish; Flow cytometry tubes; Marker pen

Equipment: Pipette; Microcentrifuge; CO_2 incubator; Fluorescence microscope; BD Accuri™ C5 Flow Cytometer

Procedure

See Fig. 10-10.

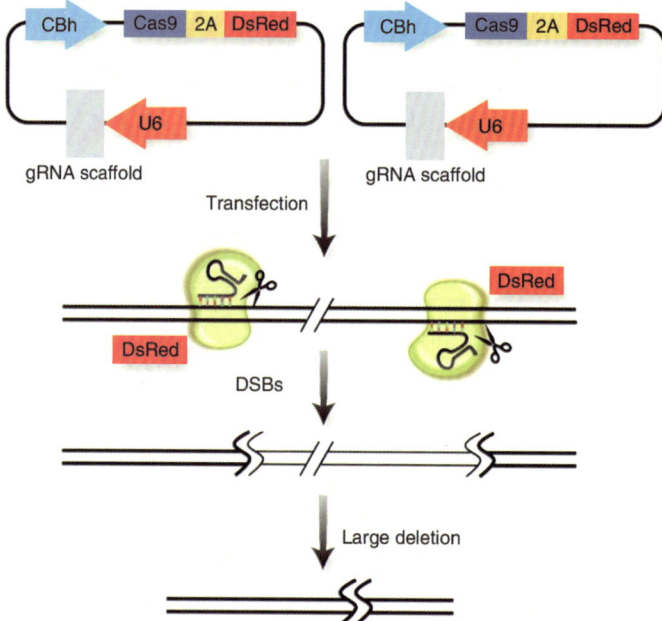

Fig. 10-10 Schematic of cotransfecting two Cas9/sgRNA coexpression vectors to induce targeted deletion in the genome

Activity Protocol

(1) The day before transfection, seed (0.5-1.0) × 10^6 HEK 293^{EGFP} cells in a 60 mm dish with 4 mL DMEM culture medium (with 10% FBS and 1% P/S). Prepare 3 dishes for each group (one negative control, two target samples).

(2) Incubate the cells at 37 ℃ in a CO_2 incubator until the cells are 50%-80% confluent. This will usually take 18-24 h, but the time will vary among cell types (optimal cell density may vary with cell type or application. Because transfection efficiency is sensitive to culture confluence, it is important to maintain a standard seeding protocol from experiment to experiment).

(3) For each 60 mm dish in a transfection, dilute 0 μg plasmid (negative control), or 6.0 μg target plasmids (3.0 μg pX459-gRNA1 + 3.0 μg pX459-gRNA2) in 250 μL Opti-MEM Medium (in a 1.5 mL microcentrifuge tube), then add 12 μL P3000™ Reagent to the tube. Mix gently and incubate at room temperature for 5 min.

(4) For each 60 mm dish in a transfection, dilute 15 μL Lipofectamine™ 3000 Re-

agent in 250 μL Opti-MEM Medium (in a 1.5 mL microcentrifuge tube). Mix gently and incubate at room temperature for 5 min.

(5) Combine diluted DNA with P3000™ Reagent (from step 3) and diluted Lipofectamine™ 3000 reagent (from step 4), mix gently and incubate at room temperature for 10-15 min.

(6) During the incubation for the formation of DNA-liposome complexes, take the cells from CO_2 incubator, discard the DMEM culture medium, and wash the cells in each dish with 2 mL Opti-MEM Medium once.

(7) For each transfection, add 0.5 mL of Opti-MEM Medium to the microcentrifuge tube containing DNA-liposome complexes. Mix gently and add the diluted complex solution to the rinsed cells (with 1 mL Opti-MEM Medium in each dish).

(8) Incubate the cells with the complexes at 37 ℃ in a CO_2 incubator for 24-48 h.

(9) After finishing transient transfections, carefully take out 60 mm dishes with transfected cells from CO_2 incubator and observe the fluorescent signal of EGFP and DsRed proteins under an inverted fluorescence microscope.

(10) Select at least three views for a 60 mm dish and take photos of the bright field image and fluorescence image for each view.

(11) For a given view of a 60 mm dish with transfected cells, count the total cell number of a bright field image and the number of cells which expressing GFP or DsRed and show fluorescent signal.

(12) For a 60 mm dish with transfected cells, calculate the transfection efficiency according to the following formula: fluorescent cells (sum of three view)/ total cells (sum of three views) × 100%.

(13) Compare the transfection efficiency of different transfection samples for each group.

(14) After taking photos of bright field and fluorescence images for each dish with transfected HEK 293^{EGFP} cells, discard the culture medium in the dishes and wash the cells with 2 mL 1 × PBS one time to remove the remaining culture medium.

(15) Add 0.5 mL 0.25% trypsin-EDTA to 60 mm culture dishes and incubate at 37 ℃ for 3 min. Then gently disperse the detached cells with a transfer pipet several times and add 0.5 mL FBS to the dish to inactivate trypsin-EDTA. Add 4 mL DMEM medium to the same dish.

(16) Transfer the cell suspension to a 15

mL centrifuge tube and collect the cells by centrifuging at room temperature at 900 rpm for 5 min. Discard the supernatant and resuspend the cells with 2 mL 1 × PBS and gently disperse the cells with a transfer pipet 10-20 times.

(17) Count the cell number and determine their viability with 0.4% trypan blue staining.

(18) Centrifuge the cell suspension at room temperature at 900 rpm for 5 min. Discard the supernatant and resuspend the cells in an appropriate volume of 1 × PBS to a density of 0.5×10^6 cell/mL.

(19) Transfer 1 mL of the cell suspension to a 5 mL polystyrene round-bottom tube with cell-strainer cap to detect the transfection efficiency of cells with the BD Accuri™ C5 Flow Cytometer.

(20) The remaining cells will be used for genomic DNA extraction to detect the effect of targeted deletion of EGFP by CRISPR/Cas9.

Section 5 PCR Analysis of Targeted Deletion in the Genome

Objective

❖ Identification of a targeted deletion in the genome by PCR analysis

Materials

Cell: HEK293^{EGFP}

Reagents: E.Z.N.A.® Tissue DNA Kit; PBS; 0.25% trypsin-EDTA; 2 × Taq Mix; Primers; 50 × TAE; SYBR Green; Agarose; SYBR Green; DNA marker (100 bp); Loading buffer

Supplies: Pipette tips; Microcentrifuge tubes; Marker pen

Equipment: Pipette; Microcentrifuge; CO_2 incubator; Gel electrophoresis apparatus

Procedure

See Fig. 10-11.

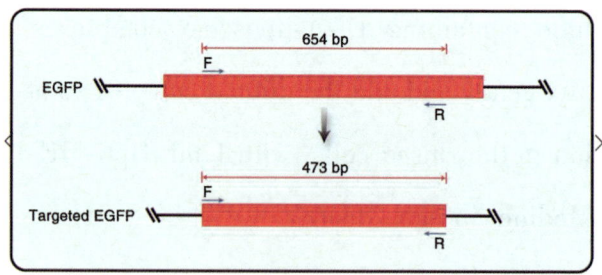

Fig. 10-11 A schematic representation of PCR used to identify targeted deletion in the genome

Activity Protocol

(1) Follow the subsequent steps to isolate genomic DNA from HEK293^{EGFP} cells transfected with paired Cas9/sgRNA co-expression vectors.

(2) Prepare the cell suspension: harvest the cells by trypsinization and then wash twice with cold PBS (4 ℃). Re-suspend cells in 200 μL PBS and proceed to Step 3.

(3) Add 25 μL OB Protease Solution. Vortex to mix thoroughly.

(4) Add 220 μL BL Buffer.

Note: *A wispy precipitate may form upon the addition of BL Buffer. This does not interfere with DNA recovery.*

(5) Incubate at 70 ℃ for 10 min. Briefly vortex the tube once during incubation.

(6) Add 220 μL 100% ethanol. Adjust the volume of ethanol required based on the amount of starting material. Vortex to mix thoroughly.

(7) Insert a HiBind® DNA Mini Column into a 2 mL collection tube.

(8) Transfer the entire sample from Step 6 to the HiBind® DNA Mini Column including any precipitates that may have formed.

(9) Centrifuge at maximum speed (⩾ 10,000 ×g) for 1 min.

(10) Discard the filtrate and re-use the collection tube.

(11) Add 500 μL HBC Buffer.

Note: *HBC Buffer must be diluted with isopropanol before use.*

(12) Centrifuge at maximum speed for 30 s.

(13) Discard the filtrate and collection tube.

(14) Insert the HiBind® DNA Mini Column into a new 2 mL Collection Tube.

(15) Add 700 μL DNA Wash Buffer.

Note: *DNA Wash Buffer must be diluted with ethanol before use.*

(16) Centrifuge at maximum speed for 30 s.

(17) Discard the filtrate and re-use the collection tube.

(18) Repeat Steps 15-17 for a second DNA Wash Buffer wash step.

(19) Centrifuge the empty HiBind® DNA Mini Column at maximum speed for 2 min to dry the column.

Note: *This step is critical for removal of trace ethanol that may interfere with downstream applications.*

(20) Transfer the HiBind® DNA Mini Column into a nuclease-free 1.5 mL microcentrifuge tube.

(21) Add 100-200 μL Elution Buffer heated to 70 ℃.

(22) Let sit at room temperature for 2 min.

(23) Centrifuge at maximum speed for 1 min.

(24) Repeat Steps 21-23 for a second elution step.

Note: *Each 200 µL elution will typically yield of 60%-70% of the DNA bound to the column. Thus two elutions will generally yield ~90%. However, increasing the elution volume will reduce the concentration of the final product. To obtain DNA at higher concentrations, elution can be carried out using 50-100 µL Elution Buffer (which slightly reduces overall DNA yield). Volumes lower than 50 µL greatly reduce yields. In some instances yields may be increased by incubating the column at 70℃ (rather than at room temperature) upon the addition of Elution Buffer.*

(25) To verify that the EGFP gene has been successfully deleted by CRISPR/Cas9, use PCR to amplify the target amplicon encompassing the two sgRNA target sites using a pair of primers and the isolated genomic DNA as a template. Set up the PCR reaction as following (Table 10-5):

Table 10-5 Requirements for setting up a genomic PCR reaction

Component	Amount (µL)
Genomic DNA	1
EGFP Forward Primer (100 nM)	1
EGFP Reverse Primer (100 nM)	1
Taq Mix, 2×	10
ddH$_2$O	7
Total	20

(26) Run the PCR reaction on the thermocycler according to the following program: initial denaturation cycle at 95℃ for 5 min, 36 cycles of denaturation at 95℃ for 30 s, annealing at 60℃ for 30 s, extension at 72℃ for 1 min, and an additional final elongation step at 72℃ for 10 min. The samples can be kept at 25℃ until the next step.

(27) Run the PCR product on a 2% agarose gel.

Section 6 Know More

Hornless cows that are unable to gore farmers or threaten dog walkers are being bred by scientists.

Animal geneticist Dr. Alison Van Eenennaam, of the University of California, discovered that it is possible to splice the "hornless" gene from Aberdeen Agnus cattle into the genome of the common black-and-white Holstein dairy cows, which results in calves born without protrusions. Instead, they grow soft hair on the parts of their heads where hard mounds normally emerge. In

British farms, horned cattle pose threat to handlers, other stock, or general public. Most dairy cows in the UK undergo a painful "dehorning" process when they are calves. However, a gene-editing program would mean that farmers would not have to carry out this process. It also makes them easier to pack into pens and trucks because apart from posing risk, horns take up space; therefore, genetic editing could potentially save the industry millions of pounds a year.

The first calves, Spotigy and Buri, were created using *in vitro* fertilization (IVF) techniques. The team at the University of California are hoping that their offspring will also be hornless even if they are bred with horned cows. If successful, it will allow the industry to bypass decades of breeding hornless, cows (Fig. 10-12).

Fig. 10-12 Horned (left) and dehorned (right) cattle. Researchers have now succeeded in editing the POLLED locus in cattle

The researchers are also hoping to perfect a technique to genetically design cattle so that they only produce male offspring, which grow faster than females, and to engineer cows which are less prone to pneumonia, and thus would require lesser antibiotics.

Taking CRISPR technology further: New CRISPR genome "prime editing" system

The prime editing system builds up from the CRISPR editing technique by heavily modifying the Cas9 protein and the guide RNA. Instead of cutting both the strands, the altered Cas9 "nicks" a single strand of the double helix. The guide RNA of the prime editing system is called the pegRNA, which contains an RNA template for the desired DNA sequence to be inserted into the genome at the target location. This process requires a second Cas9-associated protein: a reverse transcriptase enzyme that can make a new DNA strand from the RNA template and insert it at the nicked site (Fig. 10-13). Prime editing is more versatile and accurate than previously established genome editing techniques. It can directly and accurately introduce new genetic information into a specified DNA site. This technique uses the Cas9 nickase fused with a reverse transcriptase for a single-strand cut at the target site and the generation of new DNA strands. This technique allows to cut out the mutated DNA sequence and replace it with an edited strand at the target location in one step, making the deletion and insertion of sequences into the eukaryotic genome much easier. The inserted strands are not at risk of breaking and entire sections could be altered using this technique. Since two-thirds of all genetic disorders are caused by single point genetic mutations, prime editing may be used to treat 89% of the diseases caused by genetic variation[4].

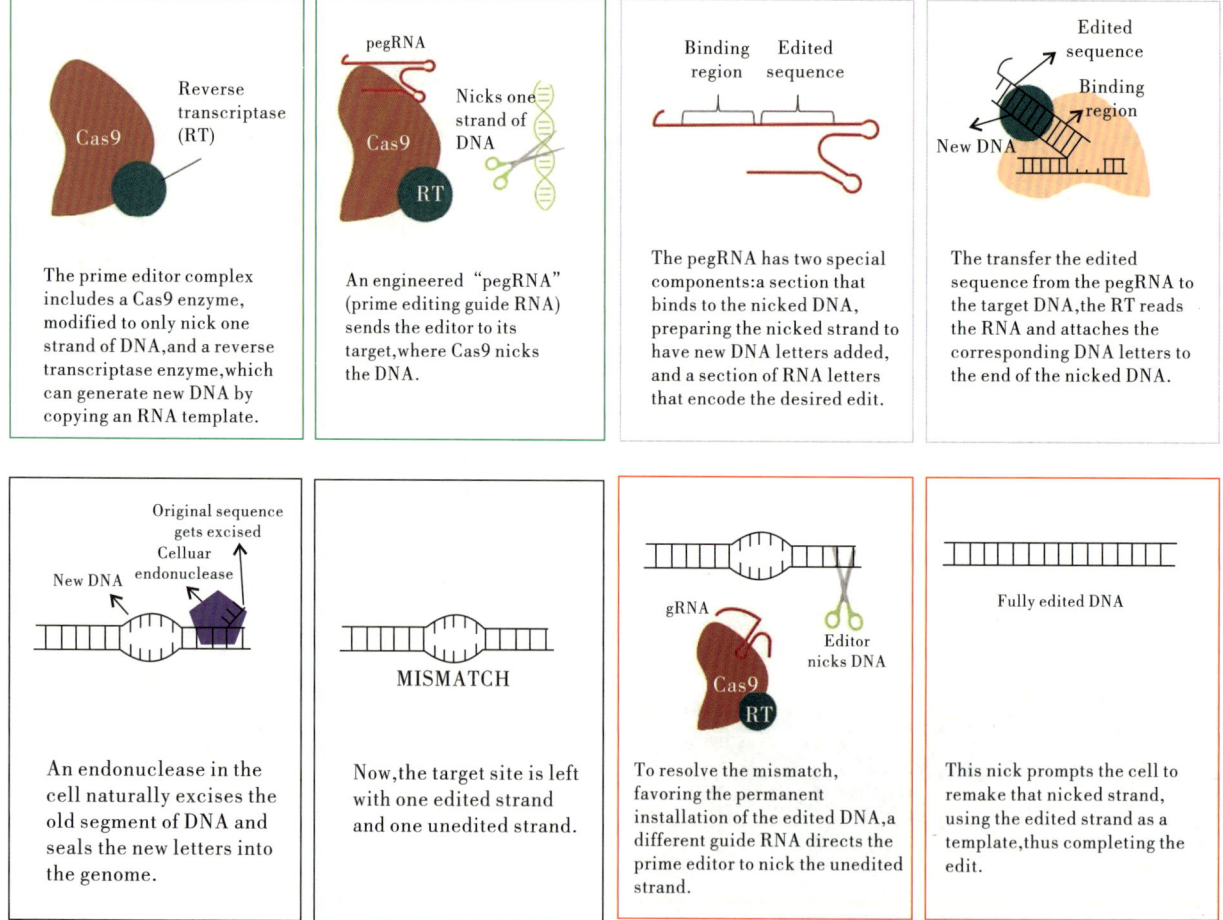

Fig. 10-13 A schematic diagram depicting prime editing procedure

What is SHERLOCK?

Unlike the CRISPR-related DNA-targeting enzymes (e.g., Cas9 and Cpf1), the CRISPR-related enzymes that target RNA (e.g., Cas13a) remain active after cutting the target RNA. They may exhibit indiscriminate cutting of non-target RNAs in a series of events known as collateral cleavage. Nucleic acid detection with Cas13a CRISPR technology was named Specific High-sensitivity Enzymatic Reporter unLOCKing (SHERLOCK). SHERLOCK could one day be used to combat viral and bacterial outbreaks, monitor antibiotic resistance, and detect cancer[5-6].

»References

[1] THURTLE-SCHMIDT D M, LO T W. Molecular biology at the cutting edge: a review on CRISPR/CAS9 gene editing for undergraduates[J]. Biochemistry and molecular biology education, 2018, 46: 195-205.

[2] KNOTT G J, DOUDNA J A. CRISPR-Cas guides the future of genetic engineering [J]. Science, 2018, 361: 866-869.

[3] SASHITAL D G. Pathogen detection in the CRISPR-Cas era[J]. Genome Med, 2018, 10: 32.

[4] COHEN, J. Prime editing promises to be a cut above CRISPR [J]. Science, 2019, 366: 406. doi: 10.1126/science.366.6464.406.

[5] GOOTENBERG J S, ABUDAYYEH O O, et al. Nucleic acid detection with CRISPR-Cas13a/C2c2 [J]. Science, 2017, 356: 438-442.

[6] GOOTENBERG J S, ABUDAYYEH O O, et al. Multiplexed and portable nucleic acid detection platform with Cas13, Cas12a, and Csm6 [J]. Science, 2018, 360: 439-444.

Postlab Focus Questions

1. What components do you need to introduce into the eukaryotic cells to induce a double-strand break (DBS) in the genome?

2. What kind of mutations can you create using CRISPR/Cas9?

3. Do you know of any advantages of using type IIS enzymes (e.g., *Bbs* I) in designing customized sticky ends?

4. Can you design a strategy to insert a copy of the GFP gene into a specific locus in the mammalian genome?

(Written by He Zuyong)

Chapter 11 Biosafety in Laboratory

A biological laboratory is an important hub for both educational and scientific research advancement. It is an important platform to cultivate and improve students' awareness and practical abilities to innovate. Since the safety of teachers and students is paramount, attention should be paid to laboratory safety and biosafety to prevent accidents and mishaps.

Section 1 General Laboratory Safety

Maintaining personal safety and that of surroundings is paramount in a biological laboratory. The laboratory houses many reagents, including highly toxic compounds, pathogenic substances, and carcinogens, etc. If you come into contact with these reagents but have not followed safe and effective protective measures, you will potentially cause great harm to yourself and other operators.

A laboratory contains many equipment such as autoclaves, drying ovens, and high-speed centrifuges, among others. Users must be trained to operate these instruments properly. The improper operation of these instruments would not only damage the equipment, but may also cause injury to the operator. In addition, the failure of some high-power instruments may cause power failure or even fire. Therefore, we should become familiar with the potential dangers in the laboratory and pay attention to laboratory safety.

Part 1. Potential hazards in the laboratory

1. Biological risk

People working in biological laboratories are at risk from exposure to pathogenic microorganisms, experimental animals, biological toxins, and other hazards. Among these, pathogenic microorganisms pose considerable risk in biological laboratories as they are highly infectious and difficult to detect. Experimental animals could also be infected by pathogens and act as carriers, thereby posing increased risk to operators. For example, the risk of infection will be increased if the skin is accidentally cut during an experiment.

2. Chemical risk

Many chemical reagents are used in the laboratory. These include common organic reagents such as ethanol and diethyl pyrocarbonate (DEPC), and corrosive reagents such as TRIzol, hydrochloric acid, and strong bases. Ethidium bromide is a highly sensitive fluorescent compound used to label DNA in agarose and polyacrylamide gels. It is also a strong mutagen and carcinogen. DEPC is an effective nuclease inhibitor that can cause strong irritation in the eyes and airway. It is also a potential carcinogen. Acrylamide is a neurotoxin, which can be absorbed through the skin and respiratory tract and can cause damage to the nervous and reproductive systems. TEMED is flammable, corrosive, and highly neurotoxic. Trichloro-methane can irritate the skin, eyes, mucous membrane, and respiratory tract. The main component of TRIzol is phenol, a strongly corrosive chemical that can damage the skin, mucosa, liver, kidney, and central nervous system. Liquid nitrogen is often used to freeze and pulverize animal tissues and to preserve biological specimens. At normal pressure, the liquid nitrogen temperature is −196 ℃ and a direct skin contact with liquid nitrogen can cause frostbite. Thus, one must be careful and take necessary protective steps when handling these reagents.

3. Ultraviolet irradiation

Ultroviolet (UV) lamps are used for surface sterilization (e.g., to sterilize ultra-clean benches) and visualization of nucleic acids (e.g., when cutting or imaging gels). Prolonged exposure to UV-rays can damage the eyes and skin. Similarly, an operator can occasionally break the glassware and injure himself/herself. Therefore, we should be careful when carrying out experiments.

4. Fire and explosion

Many inflammable and explosive chemicals are housed in laboratories. The improper usage of these chemicals may cause accidental fires or explosions. The alcohol lamp should be covered with a lid to cut off its oxygen supply after usage. It should not be blown out with mouth because mixing volatile alcohol with air can cause ignition and it can explode.

5. Electricity and water

The automatic temperature control devices of ovens, water baths, metal baths, and other heating equipment must be properly maintained in working order. If the thermostat is out of order, please turn the power off immediately. Putting inflammable and explosive reagents into these instruments is strictly prohibited. Another laboratory hazard that may affect the use of these instruments is the possibility of burst pipes or other leakages in the building. If this happens, please switch the instruments off and plug the source of the leak, if possible.

Part 2. How to ensure laboratory safety

To ensure the safety of operators and the environment, we should always proceed according to the following instructions:

1. Safety awareness

The main reason behind accidents is the lack of safety consciousness. Therefore, it is necessary to receive proper laboratory safety training before working in the laboratory for the first time. This training includes relevant policies and regulations, knowledge of biosafety and the pathogenic hazards, standardized operating procedures for experimental activities, instruments and equipment, biosafety practices, and emergency handling methods for unexpected events.

In addition, when entering the laboratory, the user should first become familiar with the laboratory environment, the reagents, and the equipment that he/she will be using. Following standardized operation procedures and personal protection should be non-negotiable.

2. Standardization of instrument usage

Autoclaves, ovens, ultrasonic breakers, electrophoresis apparatus, and high-speed centrifuges are commonly used by technicians, students, and researchers. If the standard operation procedures (SOPs) of instrument usage are not followed, accidents may happen. Therefore, students must be trained by experienced staff before they are allowed to use the instruments and equipment for the first time. Students should operate the instrument after being fully trained with the appropriate operation procedure for each instrument. For instance, pressure must be released once you have finished using a pressure vessel; samples must be balanced properly prior to using a centrifuge; each sample must be stationary before using an oscillator; during electrophoresis, the power supply should be turned off before the gel is removed from the buffer to prevent electrical shock, and so on. Pay close attention to the state of the instrument during operation. If there are any abnormalities, the operation must be stopped as soon as possible.

3. Standard disposal of hazardous waste

Various solid, liquid, and gaseous waste products are often produced during experiments. If not handled properly, they can contaminate and damage the laboratory and surrounding environment, thereby potentially exposing laboratory personnel to significant risk. Therefore, the experimental wastes should be segregated into correctly labeled specific containers according to the type of waste, and commission qualified companies to regularly collect the waste for proper disposal. Biological waste, such as bacterial cultures, used cell culture medium, and discarded biological specimens, must be sterilized at high temperatures prior to disposal. It should be prohibited to mix biologic-al waste with household waste or discharge biological waste directly into the sewer.

Part 3. Notes for laboratory operation

Although there are potential safety risks when performing experiments, accidents can be prevented as long as: standardized operation procedures are in place, equipment is maintained in safe working order, and the correct protective measures have been taken.

(1) When entering the laboratory, you must wear a laboratory coat and gloves that completely cover the hand and wrist. Wear a mask, safety glasses, or goggles, where necessary. When leaving the laboratory, these must be taken off and kept in the laboratory area. Do not wear laboratory coats or gloves when you are in the living or office areas.

(2) Researchers should take extra care when handling sharp instruments to avoid cutting themselves. If one has cut the skin inadvertently, disinfect the wound and apply a bandage or band-aid, and wear double gloves.

(3) Reagents should be returned after use, and any spillage should be immediately cleaned up. Any body parts exposed to contaminants/reagents should be rinsed promptly.

(4) Eating and smoking are prohibited in the laboratory. Personal belongings are not allowed on the laboratory bench.

(5) When using dangerous chemicals and equipment, be sure to follow the operating

procedures.

(6) The use of volatile and irritating agents must be carried out in a fume hood.

(7) When using liquid nitrogen, you must wear thick cotton gloves or insulated gloves.

(8) Do not touch electrical appliances or equipment when hands are wet.

(9) Clearly label hazardous waste according to classification and deposit the waste into special collection containers for disposal.

(10) Gas cylinders should be stored separately in the cylinder cabinet or fixed in place by a chain.

(11) After an experiment is finished, please close and sanitize the instrument. Always keep the experiment bench clean.

(12) Wash hands thoroughly using soap or hand sanitizer after each experiment and rinse them under running water for at least 20 s.

(13) When you are the last person to leave the laboratory for the day, you should turn off the electricity, water, and lights, and close the doors and windows.

(14) Inspect and maintain instruments, electrical circuits, water pipes, ventilation, and fire protection equipment. Identify problems early and deal with them promptly.

Section 2 Biosafety

Laboratory biosafety refers to the comprehensive measures taken to prevent exposing laboratory personnel to potential dangers in a laboratory setup, and to prevent the spread of these dangers outside the laboratory. With the rapid development of modern biotechnology, biosafety has become one of the major global public health issues. Laboratory biosafety is not only limited to the protection of laboratory staff, but also that of other people and the environment, which could be affected in the event of an accident.

Biosafety is an important part of national security. After the 2003 SARS outbreak, the country formulated the mandatory "general requirements for laboratory biosafety" standards, which improved biosafety awareness among laboratory staff. The country plans to integrate biosafety into the national security system to comprehensively improve its biosafety governance capacity. Only by accelerating the implementation of biosafety laws can we further safeguard national security and ensure our people's safety and health.

Laboratory biosafety and individual protection should be the foremost considerations when designing and constructing new laboratory facilities and standardized operating procedures, to ensure that laboratory staff and their working environment are protected from biological hazards. Laboratory biosafety is important for the stable development of universities and the safety of their staff and students. As life science scientists, we should be familiar with the laws, regulations, and standards related to biosafety management. Furthermore, we should also carry out experiments according to the operation specifications of the laboratory to ensure that we are protected from potential biohazards and accidents.

Part 1. Classification of biohazards

Hazardous items are classified into four levels according to the degree of harm they can cause:

Hazard rating I (low harm to individuals and groups)

Biological factors such as bacteria, fungi, viruses, and parasites that are not pathogenic.

Hazard rating II (moderate harm to individuals and limited harm to groups)

Microbes that can be pathogenic in humans or animals, but generally do not cause serious harm to healthy people, livestock, or the environment. Laboratory infections that do not cause serious illness, have effective treatment and prevention regimes, and with limited risk of transmission are listed in this category.

Hazard rating III (high harm to individual and limited harm to groups)

Pathogens that can cause serious diseases in humans or animals, or cause serious economic losses, but are not likely to be transmitted from one person to another by occasional contact or can be treated with antibiotics or other anti-parasitic drugs.

Hazard rating IV (high harm to individuals and groups)

A pathogen that causes very serious disease in humans or animals that is generally incurable and easily transmitted from person to person, animal to person, person to animal, or animal to animal even by accidental contact.

Part 2. Biosafety classification levels

The document "General Requirements for Laboratory Biosafety" defines four ascending levels of containment, referred to as biosafety levels 1 through 4. It describes the microbiological practices, safety equipment, and facility safeguards for the risk level associated with the handling of particular agents.

Biosafety Level 1 (BSL-1)

BSL-1 is the basic level of protection

common to most research and clinical laboratories and is appropriate for non-pathogenic agents.

Biosafety Level 2 (BSL-2)

BSL-2 is appropriate for moderate-risk agents that may cause human diseases of varying severity via ingestion or through cutaneous or mucous membrane exposure. Most cell culture labs should be at east BSL-2; however, the exact requirements depend upon the cell lines used and the type of work performed in different laboratory.

Biosafety Level 3 (BSL-3)

BSL-3 is appropriate for indigenous or exotic agents with known potential for aerosol transmission, and for agents that may cause serious and potentially lethal infections.

Biosafety Level 4 (BSL-4)

BSL-4 is appropriate for exotic agents that pose a high risk, like a life-threatening disease by infectious aerosols, to individuals, and for which no treatment is available. Laboratories dealing with such agents must be BSL-4 certified.

Part 3. Laboratory biosafety protection equipment

Using biosafety protection equipment is an effective way to protect both laboratory staff and their environment. The biosafety protective equipment commonly used in laboratories include laminar flow hoods, fume hoods, autoclaves, and emergency showers.

1. Biosafety cabinet classification

The biosafety cabinet provides an aseptic work area that protects the user from infectious splashes or aerosols generated during microbiological procedures. Three types of biosafety cabinet, designated Class I, II, and III, have been developed to meet various research and clinical needs.

Class I biosafety cabinets offer significant levels of protection to laboratory personnel and the environment when used in combination with good microbiological techniques, but they do not protect the experimental subject or cell culture from contamination. They are similar in design to chemical fume hoods. Class II biosafety cabinets are designed for work involving BSL-1, 2, and 3 materials, and they also provide an aseptic environment necessary for cell culture experiments. A Class II biosafety cabinets should be used for handling potentially hazardous materials (e.g., primate-derived and virally-infected cultures, radioisotopes, carcinogens, and toxins). Class III biosafety cabinets are air-tight and provide the highest level of protection for personnel and their environment. A Class III biosafety cabinet is required for work involving known human pathogens and other BSL-4 materials.

Biosafety cabinets protect the working environment from dust and other airborne contaminants by maintaining a constant, unidirectional flow of HEPA-filtered air over the work area. The flow can be horizontal, blowing parallel to the work surface, or vertical, blowing from the top of the cabinet onto the work surface. Depending on its design, a horizontal flow hood protects the sample (if the air flows toward the user) or the user (if the air is drawn in through the front of the cabinet by the negative air pressure inside). However, vertical flow hoods provide significant protection to the user and the sample.

2. Ultra-Clean Benches

Horizontal laminar flow or vertical laminar flow "clean benches" are not biosafety cabinets. These equipment discharge HEPA-filtered air from the back of the cabinet across the work surface toward the user, and they may expose the user to potentially hazardous materials. These devices only protect the product. Clean benches can be used for certain clean activities such as the dust-free assembly of sterile equipment or electronic devices. They should never be used when handling cell culture materials or drug formulations, or when manipulating potentially infectious materials.

3. Fume hoods

Fume hoods protect the operators and laboratory environment by safely discharging toxic, harmful, and volatile agents from the cabinet. The fume hood structure is similar to a BSL-1 cabinet but without HEPA filters. Although the fume hood expels most of the harmful gases, there would be a small amount of leakage as it is not completely sealed. Gloves, masks, and goggles should be worn when working at the fume hood.

4. Autoclave sterilizers

Autoclaves are essential for decontamination and sterilization processes. Autoclave sterilization uses high temperature (121 ℃) and pressure to create steam, which acts as the sterilization agent. The steam can be used to sterilize heat-resistant glassware, plastic labware, and certain reusable instruments. The high temperature reduces the sterilization time needed to kill microorganisms. Pressurized saturated steam (moist heat) is the most widely used form of sterilization of solids, liquids, and hollow items. The safe steaming machines come in free-standing or benchtop styles, and a choice of manually-controlled analog and electronically-controlled models.

When placing materials inside, the items

should not be too large or too tightly packed, so that it does not hinder the pass-age of steam and affect the sterilization effect. Do not block the vent hole of the safety valve, and enough space must be left so that the pressure can be released. The safety valve will not work if the vent hole is blocked, which can easily cause accidents.

5. Emergency eyewash and showers

Emergency eyewash and shower equipment rinse contaminants from the eyes and body to help prevent injuries or permanent damage. These are used in laboratories, chemical plants, paper mills, and other environments that contain potentially harmful chemicals or materials. Combination eyewash stations are all-in-one eyewash and shower units. Drench hoses serve as fixed eye-rinse stations or can be removed from their holder to direct the spray where needed. Emergency eyewash stations can either be units connected to plumbing lines or self-contained units with a built-in wash tank, or there can be personal eyewash stations and bottles. Drench showers can be mounted to the ceiling, wall, or floor to provide an emergency shower.

Emergency eyewash and shower stations should be installed close to the laboratory work area, preferably reachable within 10 walking seconds to avoid delay in treatment. However, these devices are only for preliminary treatment. In case of serious mishap, further medic-al treatment should be sought as soon as poss-ible.

Part 4. Personal protective equipment

Personal protective equipment (PPE) are used to protect laboratory personnel from physical, chemical, biological, and other harmful factors. In a biological laboratory, these devices provide the primary physical barrier to protect laboratory personnel from exposure to biohazards (aerosols, sprays, accidental inoculations, etc.).

1. Laboratory coats, isolation suits, jumpsuits

Ordinary laboratory coats can be worn in laboratories with biosafety I and II levels. Laboratory coats should be fully buttoned. Isolation suits should have long sleeves with an opening at the back. The jumpsuit offer better protection than an ordinary laboratory coat. Therefore, it is more suitable for use in microbiology laboratories and biosafety cabinets. Isolation suits with reinforced fabric at the back can be used for biosafety levels III and IV works. Laboratory coats, isolation suits, and jumpsuits must not be worn outside of the laboratory area.

2. Goggles and safety glasses

Choose the eye protection equipment

according to the operation, protect the eyes and face from potential splashes from the experimental objects. Safety glasses, even with a shield on the sides, do not provide adequate protection against splashes. Goggles should be worn on the outside of regular vision correction lenses or contact lenses to provide protection against splatter. Goggles and safety glasses must not be worn outside of the laboratory area.

3. Masks, face masks, and gas masks

A mask can protect part of the face from biohazards such as blood, body fluids, secretions, and fecal aerosols. However, it is suitable only for use in BSL-1 or BSL-2 laboratories because it cannot provide respiratory protection for the user. The face mask is made of shatter-proof plastic, which matches the shape of the face, and can be worn with a headband or hat to cover the whole face. When performing highly dangerous operations, such as cleaning up spills of infectious materials, please use gas masks for protection. Please follow the advice of qualified personnel when choosing the gas mask. Masks, face masks, and gas masks must not be worn outside of the laboratory area.

4. Gloves

Disposable latex, vinyl, or polyacrylonitrile surgical gloves should be used for general laboratory work and for handling infectious materials, blood, and body fluids. Gloves should be removed, and hands should be thoroughly washed after handling infectious substances and before leaving the laboratory. Used disposable gloves should be discarded along with infectious material in the laboratory waste bin. Stainless steel mesh gloves should be worn in cases of possible exposure to sharp instruments, such as when carrying out an autopsy. But such gloves can only prevent cutting injuries, not needle injuries. Gloves must not be worn outside of the laboratory area.

Part 5. Biosecurity emergency management

In case of an accident, laboratory personnel should immediately carry out the emergency response procedure and report to the laboratory director.

1. Stabs, cuts, or bruises

The injured should stop working immediately, take off the protective clothing, rinse the injured area with soap and plenty of running water, disinfect the wound with 70% ethanol or other skin disinfectants, and see a doctor. The cause of the injury and related microorganisms should be documented. A complete and appropriate medical record should be maintained.

2. Splash contamination

The infected area should be rinsed immediately with an eyewash for at least 15 min. Please see a doctor if necessary. If it is a small-scale splashing accident, disinfect the affected area immediately. Cover the affected area with 0.5% sodium hypochlorite or peroxyacetic acid-soaked cloth for at least 1 h. When it is a large-scale splashing accident, the laboratory supervisor and biosafety director should be immediately called to the scene of the accident to investigate the situation, determine the disinfecting procedures, and immediately report to the institute or school.

3. Skin contamination

Rinse the contaminated area with soap and water, and soak with an appropriate disinfectant such as 70% ethanol, iodine, or other skin disinfectants.

4. Mucosa contamination

Rinse the contaminated area thoroughly with plenty of running water or saline buffer.

5. Ingestion of potentially infectious substances

Take off protective clothing and immediately seek medical treatment. Report the ingested materials and incident in detail, and maintain a complete and appropriate medical record.

6. Release of potentially hazardous aerosols

All personnel must evacuate the area immediately, and any exposed personnel should receive medical treatment. The laboratory director and the biosecurity staff should be notified immediately. Access to the laboratory should be prohibited for a certain period of time so that the aerosol can disperse, and the larger particles can settle. If there is no central ventilation system, access to the laboratory should be further delayed. "No entry" signs should be posted and appropriate protective clothing and respiratory protection equipment should be worn when removing the contaminants under the direction of the biosecurity staff.

7. Rupture of centrifuge tube containing potentially infectious agents

If a centrifuge tube breaks or is suspected of breaking during centrifugation, stop the machine immediately and seal it so that the aerosol can settle. If the centrifuge breaks after it has been stopped, the lid should be closed immediately and sealed for appropriate duration. You should notify biosecurity staff if these incidents occur. All subsequent operations should be carried out with gloves (e.g., heavy rubber gloves), covered by appropriate disposable gloves on the outside if necessary. Tweezers should be used to remove glass fragments. All broken tubes, glass fragments, centrifugal barrels, cross shafts, and swivel heads should be immersed in a non-corrosive disinfectant effective to kill the contaminants. The centrifuge chamber should be cleaned several times with the an appropriate disinfectant, wiped with wet cloth, and dried. All materials used in cleaning should be treated as infectious waste. If necessary, all sealed centrifugal drums should be loaded and unloaded in a biosafety cabinet.

8. Leakage of highly pathogenic microorganisms

Close contaminated laboratories or areas and sterilize the site. Operators should be isolated for observation and treatment. Close contacts should be placed under medical observation. Animals infected with or suspected of being infected with the epidemic should be isolated or culled.

(Written by Chen Xiaoxia, Zhang Yan)

Chapter 12 The Laboratory Report

Writing laboratory report provides students with an opportunity to develop a skill that will be extremely valuable in their academic careers. The laboratory report describes what was done in the class, the raw data together with the analysis from the student, and the conclusion based on this analysis. Importantly, merely recording the observed results is not sufficient; a good laboratory report demonstrates the student's comprehension of the concepts behind the data and improve the writer's understanding of the principles of the experiment. Therefore, it is essential to train and to practice the skill of organizing ideas carefully and expressing them coherently.

The Format of the Laboratory Report

While there is no single perfect format for laboratory reports, the general format of a laboratory report includes sections that describe the background, materials and procedures, results, and conclusions (Fig. 12-1). In general, reports should be divided into the following sections, in order of their appearance, and each section of the report should be headlined and arranged in an appropriate, easy-to-understand sequence[1].

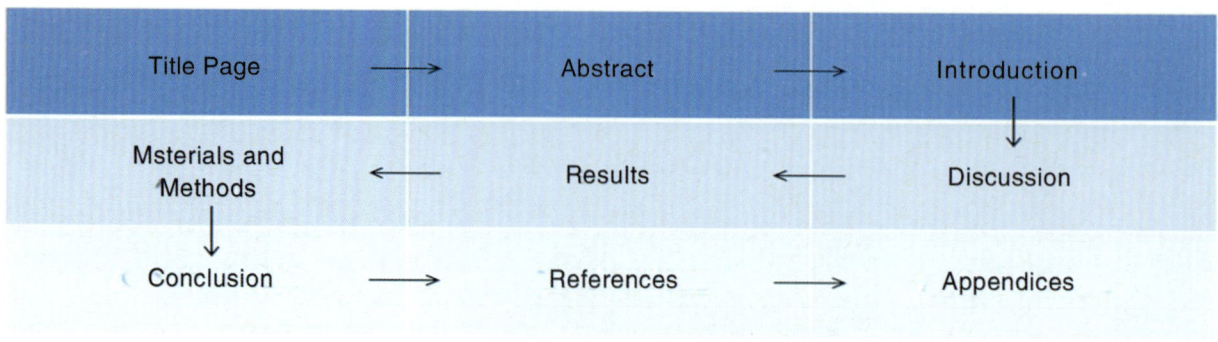

Fig. 12-1 Sections of the laboratory report

(1) The title page contains the name of the experiment, partners, teachers, and the date(s) on which the experiment was performed. Titles should be straightforward and informative (i.e. Not "Experiment #1" but "Experiment #1: Isolation of genomic DNA").

(2) The abstract summarizes five essential aspects of the report: the purpose of the experiment, methodology, key findings, significance, and major conclusions. The abstract may also include a brief reference to the theory or methodology. The abstract should be one paragraph of 100-200 words with no references and graphs.

(3) The introduction provides the background theory and scientific literature available in the field. The objective of this section should state the problem that your procedure and data attempts to answer, and should inform the reader precisely why the project was undertaken. A concise description of the relevant theory should be provided, and subheadings could be used if there is a lot of introductory material, such as "Bacteria transformation" or "Blue/white screening".

Information Box 12-1: Focus the introduction with following points:

(1) Summarize findings from previous studies with appropriate references.

(2) Catch the reader's attention with a pressing issue.

(3) State your hypothesis (e.g. Why do you care about these questions? What do you predict to happen?).

(4) The materials and methods can tell readers what equipment and materials were used in the experiment. Experimental procedure provides a clear, and accurate schematic drawing of the experimental setup, showing all interconnections and interrelationships. All the information needed for a reader to duplicate the experiment independently should be included in this section. Sometimes you can document occasions when you have deviated from established protocol (e.g., "At step 5 we performed four repetitions instead of three"). Special procedures used to ensure specific experimental conditions, or to maintain a desired accuracy in the information obtained should be described. Copying operating procedures from other textbook or website would be inconsistent with the work completed in the class and is not acceptable.

(5) The results are consisted of raw data, figures and tables. In this section, the experimental data should be presented explicitly and all numerical data should be tabulated.

Figures and graphs are picturesque to

exhibit your data. Note that all the figures and graphs should be clear, easy to read, and well-labeled with informative figure legends such as experimental strategy or experiment details. Tables may be presented in any format that clearly differentiates between rows and columns. Headings should indicate the units associated with any value. All of figures, tables and additional data should be numbered sequentially in the order that it is referenced in the written text.

(6) The discussion is the section used to interpret data. This is the most important part of your report because here you exhibit your ability to understand the experiment beyond the simple level of completing it. Discuss possible significance, trends, or correlations within your results. Explain whether or not you accept your hypothesis and why you have come to this conclusion using your analyzed data as support. Also, discuss potential sources of error in your experiment and how they could have affected your results. This part of the report focuses on understanding "what is the significance or the meaning of the results?"

Information Box 12-2: The following aspects should be discussed to illuminate the significance or the meaning of the results.

(1) What have you found? What is the significance of your results? Compare expected results with those obtained from your experiment.

(2) What do the results indicate? Compare your results to similar investigations, and explain what you know based on your results and draw conclusions.

(3) What questions might you raise? Find appropriate explanations for problems in the results.

(4) Analyze experimental error and analyze the strengths and limitations of your experimental design or operation.

(7) The conclusions state what you now know for sure. This section should answer the question: "So what?" Explain any broader implications of your experiment. Suggest further research that could be done to expand on your work. The conclusion might be a place to discuss weaknesses in the experimental design; the future works that need to be done to verify or improve your conclusions.

(8) The references include your lab manual and any outside reading you have done. Using standard bibliographic format, cite all published sources that you have consulted during the experiment and the preparation of your laboratory report. List the author(s), the title of paper or book, the name of journ-

al or publisher as appropriate, page number(s) if appropriate, and the date. If a source is included in the reference list, it must also be referred to at the appropriate place(s) in the report.

(9) The appendices typically include raw data, calculations, graphs, pictures, or tables that have not been included in the report itself. Each item should be contained in a separate appendix. Make sure that you have referred to each appendix at least once in your report.

≫ References

https://advice.writing.utoronto.ca/types-of-writing/lab-report/.

(Written by Zhang Yuchan, Zhang Yan)

Appendix

Section 1 Experimental Instruments

Life science advances go hand in hand with the development of novel experimental technologies and instruments. Keeping up-to-date with these technologies and instruments is, therefore, very important for laboratory research in universities and other research institutions. This section will introduce the basic principles, operation steps, and precautions of some of the most commonly used instruments. Proficiency in the operation of related instruments is an important skill that will help you in carrying out your experiments.

1. Micropipettes

A micropipette is used to transfer small amounts (< 1 mL) of liquids. The scales on micropipettes are in microliters (1000 μL = 1 mL). They come in several sizes, which can pipette different volume ranges: P2 = 0.2-2 μL, P10 = 1-10 μL, P20 = 2-20 μL, P200 = 20-200 μL, and P1000 = 100-1000 μL. They are used in conjunction with disposable (often sterile) plastic tips. The smaller micropipettes (P2 and P10) use the white tips; P20 and P200 use the yellow tips; the P1000 pipettes use the larger blue tips (Fig. A-1). The following is an illustration of a micropipette:

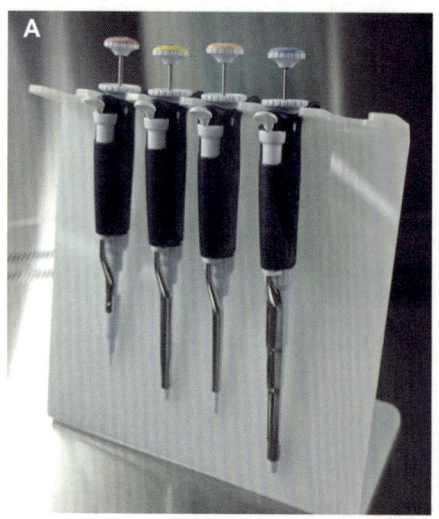
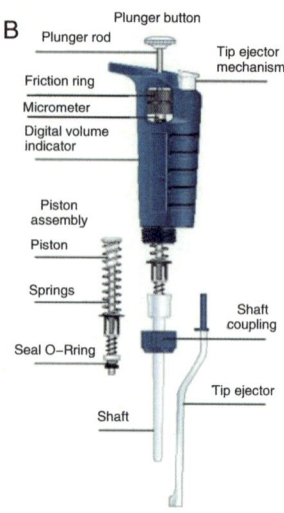

Fig. A-1 The structure of a micropipette
A. micropippettes; B. parts of a micropippette

Precautions

(1) Never exceed the upper or lower limits of micropipettes. They are expensive and delicate instruments which we cannot afford to damage.

(2) Set the desired volume by turning the centrally located rings clockwise to increase volume or counterclockwise to decrease volume (Fig. A-2). Some examples are provided below:

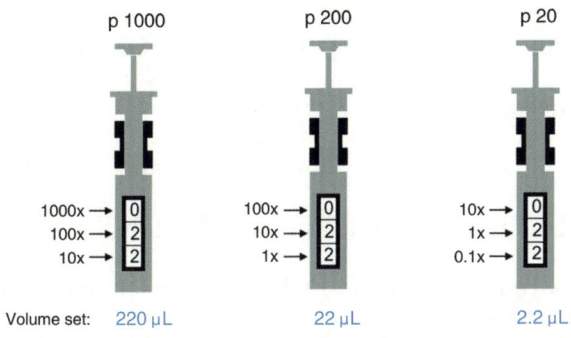

Fig. A-2 How to adjust volumes on a micropipette

(3) Place a tip at the discharge end of the pipette.

Note: If sterile conditions are required, do not allow the pipette tip to touch any object (including your hands).

(4) The plunger will stop at two different positions when it is depressed. The first of these stopping points is the point of initial resistance and is the level at which you should press to obtain the desired solution volume. Because this first stopping point is dependent on the volume that is being transferred, the distance you have to push the plunger to reach this point will change depending on the volume being pipetted. The second stopping point can be found when the plunger is depressed beyond the initial resistance until it is in contact with the body of the pipette. At this point the plunger cannot be depressed any further. This second stopping point is used for the complete discharge of solutions from the plastic tip. This second stop should only be reached when expelling the liquid, not when drawing it into the pipette. Before continuing, practice depressing the plunger to each of these stopping points until you can easily distinguish between the two.

(5) Depress the plunger until you feel the initial resistance and insert the tip into the solution, just barely below the surface of the liquid and not as deep as possible.

(6) Carefully and slowly release the plunger.

Note: If the solution you are pipetting is viscous, allow the pipette tip to fill to the final volume before removing it from solution to avoid bubble-formation in the plastic tip, which will result in an inaccurate volume. If you release the plunger too quickly, it will suck liquid up into the pipette and damage it.

(7) Discharge the solution into the appropriate container by depressing the plunger. This time, depress the plunger to the point of initial resistance, wait one second, and then continue pressing the plunger as far as it would go to discharge the entire volume of solution.

(8) Remove tip by pressing down on the tip discarding button.

2. Thermal cycler

The thermal cycler, also known as a PCR machine or DNA amplifier, is a common laboratory apparatus used to amplify segments of DNA via PCR. Thermal cyclers may also be used in laboratories to facilitate other temperature-sensitive reactions, including restriction enzyme digestion. The device has a thermal block with slots where tubes containing the reaction mixtures are be inserted. The machine can then increase or decrease the temperature of the block in discrete pre-programmed steps (Fig. A-3).

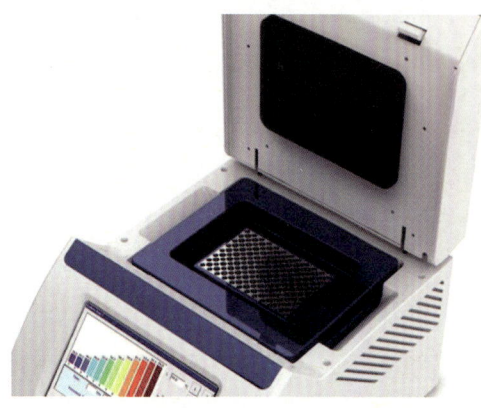

Fig. A-3　Thermal cycler

Operation steps

(1) Open the lid, place the reaction tubes into the sample holes, and close the lid.

(2) Start the instrument and wait for it to self-check before entering the document management interface.

(3) Open or create a folder: click OPEN or NEW to open, or create a folder.

(4) Open or create a file: click OPEN or NEW to open or create a file.

(5) Set up each step of the PCR program: select a STEP and click EDIT to edit the current step.

(6) Input the temperature and time for denaturation, annealing, and extension steps as required for the PCR. Input the number of cycles at CYCLE for each STEP, and then click go to select the starting step for the next cycle.

(7) Click OK to complete the PCR program setup.

(8) After the program setup has been completed, click SAVE to create a new file, and then click OK to save the file.

(9) Open the folder and select the file that you want to run. Click RUN.

(10) At the start of the operation, the interface displays the current temperature, the number of completed cycles, the remaining time, and other information.

(11) After the PCR program has finished, remove the reaction tubes, and shut down the instrument.

Precautions

(1) The PCR instrument should be used preferably in an air-conditioned room.

(2) The instrument lid cannot be opened when the reaction is in progress.

(3) After the PCR cycle have been completed, the tubes should be taken out or kept at an appropriate temperature, e. g., 15 ℃. This temperature should not be too low, and the time that it is stored at this temperature should not be too long.

(4) After the product is taken out, the machine should be covered to prevent dust from accumulating on the machine, which would affect temperature detection and control.

(5) After the PCR cycles have been completed, the machine should be turned off, because the machine will be damaged if the hot cover is left on for too long.

3. High-speed refrigerated centrifuge

High-speed refrigerated centrifuge can reach speeds of 30,000 rpm and can cool the sample at the same. The high-speed rotating head produces a strong centrifugal force, which allows the substances in suspension or emulsion to settle depending on their settling coefficient. Thus, samples can be separated, concentrated, or purified using this technique. The refrigeration allows for the accurate temperature control inside the centrifuge. Not only does this maintain the integrity of biological reagents that require low-temperature storage, but also improve the centrifugal efficiency.

The rotor is the main part of the centrifuge. A multi-function centrifuge can be equipped with several different types of rotors. Rotor selection is based on the size of the sample to be centrifuged (volume and mass), the maximum speed that the rotor can achieve, and the centrifuge tube capacity. The rotor is usually made from alloys of titanium or aluminum. Specialized centrifuge tubes made from hard plastic with polyethylene lids are used to collect pellets of nucleic acids, microorganisms, cell fragments, cells, large organelles, sulfuric acid deposits, and immune deposits.

Operation steps

(1) Turn on the power switch.

(2) Press the OPEN button to open the lid. Clean the cavity.

(3) Select the appropriate rotor and place it on the centrifugal shaft, then gently lift the rotor to check whether it has been

installed properly.

(4) Parameter setting: press the SPEED and TIME buttons to enter the required speed and time, respectively. The speed can be changed between rpm and × g. Press TEMP to enter the required temperature.

(5) Balance the centrifuge tube containing the samples, and put these into the slots inside the rotor. Screw the rotor cover close.

(6) Press the START to start centrifugation.

(7) When the centrifugation is complete, the instrument will stop automatically. You can also manually stop centrifugation by pressing the STOP. Only when the centrifugal speed has reduced to "0" can the lid be opened.

(8) After removing the centrifuge tubes, clean the cavity with a dry cloth. Leave the lid open until the centrifuge is dry, and then close it.

(9) Turn off the power switch after use.

Precautions

(1) The centrifuge should always be placed on a flat surface.

(2) Before starting the machine up, check whether there are any contaminants inside the cavity.

(3) Select the appropriate rotor for the centrifuge and assemble it accurately.

(4) The centrifuge tubes and their contents must be balanced and should be loaded symmetrically.

(5) The centrifuge tubes should not be overfilled.

(6) Before starting, be sure to screw the rotor cover close and close the lid to the centrifuge.

(7) Do not open the lid if the centrifuge has not completely stopped.

(8) The centrifuge should be stopped immediately if it shakes or makes abnormal noises.

(9) Clean the inside of the centrifuge regularly.

4. Inverted fluorescence microscope

A fluorescence microscope is an optical microscope that uses fluorescence and phosphorescence to study the properties of materials. An inverted fluorescence microscope has an objective lens, condenser, and an inverted light source. An excitatory light shines through the objective lens to illuminate the specimen, which emits light. Light from the specimen passes back through the objective lens and is separated from emission light by a two-color beam barrier filter. The working distance between the objective lens and the condenser lens is very large. Therefore, you can directly study objects in a petri dish

(Fig. A-4).

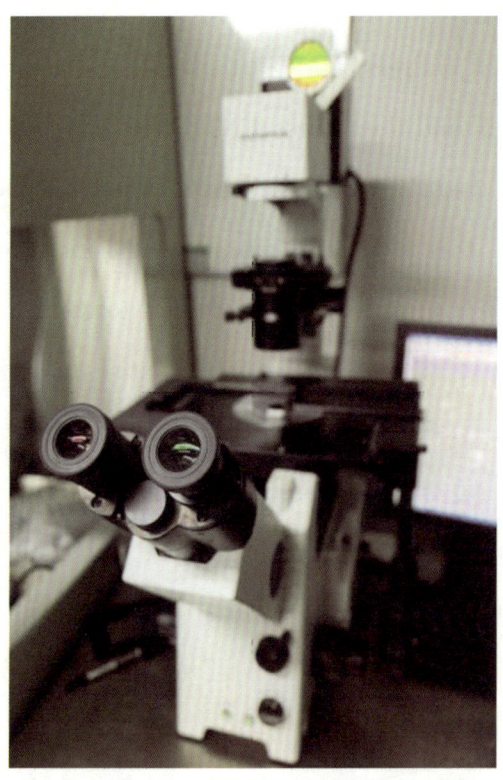

Fig. A-4 Inverted fluorescence microscope

Operation steps

(1) Turn on the microscope, the power supply for the mercury lamp, and the computer.

(2) Place the specimen on the stage.

(3) Select the appropriate objective lens and adjust the brightness of the light source to a moderate level.

(4) Rotate the control knobs for the stage, manually adjust the coarse focus knob, and look at the specimen under the eyepiece.

(5) Adjust the white balance on the image acquisition software before locating the specimen.

(6) After locating the specimen, manually fine focus until the image is clear.

(7) Press the image capture key on the software interface to obtain the specimen image.

(8) Turn off the halogen lamp or turn it down as much as possible.

(9) Select the appropriate fluorescent excitation filter cube and open the corresponding shutter.

(10) Select the appropriate neutral density (ND) filter and adjust the aperture to improve the contrast.

(11) Adjust the exposure time and gain on the software interface, and manually adjust the focus until the image is clear.

(12) Press the image capture key on the software interface and save the file.

(13) After you have finished capturing your images, withdraw the fluorescence excitation filter cube from the microscope.

(14) Turn off the microscope and mercury lamp.

Precautions

(1) Please leave the mercury lamp on for 30 min before turning it off. Once turned off, the lamp cannot be turned on for at least 1 h to maintain the life of the mercury lamp.

(2) Before turning off the microscope, minimize the brightness of the light source.

(3) If there are any contaminants on the objective lens and stage, wipe them off immediately.

(4) Before locating the specimen, adjust the white balance in blank areas where there is no specimen and perform off-color calibration.

(5) Keep the microscope dry, clean and dust-free, and clear of any chemical contamination.

5. Sonicator

A sonicator is an instrument that uses sound to produce a cavitation effect in liquid. The machine itself is made up of a power converter, a probe, and a soundproof box. The power converter converts 50/60 Hz mains voltage into high-frequency electrical energy, which is changed into mechanical vibration via the piezoelectric transducer. It can be used to break up and isolate cells, tissues, viruses, bacteria, and various inorganic materials. Sonication can be used in the following processes: emulsification, separation, homogenization, extraction, cleaning, and the acceleration of chemical reactions.

Operation steps

(1) Turn on the machine.

(2) Press the AMPL to set to maximum intensity, and then press ENTER to confirm.

(3) Press TIMER to set the sonication time, and then press ENTER to confirm.

(4) Press PULSER to set the cycle time that the sonicator vibrates (ON) and stops (OFF), and then press ENTER to confirm.

(5) Clean the probe with deionized water and dry it.

(6) Dip the probe into the sample (to a depth of 5 cm). If a microtube is used, the probe can be inserted to a depth of 1 cm.

(7) Press the START to start the sonicator.

(8) After sonication has finished, clean the probe with deionized water and dry it.

(9) Turn off the machine.

Precautions

(1) Attention should be paid to the use of the micro-probe, the amplitude should not exceed 40%, and the micro-probe should not be allowed to vibrate in the air for more than 10 s.

(2) The probe should be immersed deep enough to prevent air from being injected into the sample, gas dispersal, and foam formation.

(3) The probe should be properly maintained. If the probe is eroded, the power output will be reduced.

6. Microplate spectrophotometer/reader

A microplate reader is a spectrophotometer equipped with a built-in monochromator, which measures multiple wavelengths simultaneously. It can be used in conventional end-point and dynamic kinetics studies, and spectral and pore scanning. This instrument can detect wavelengths between 200-999 nm and can be used for rapid 260/280 ratio detection, quantitation of DNA, RNA, and protein samples; and to monitor cell growth in microporous plates (Fig. A-5).

Fig. A-5 Multi-volume spectrophotometer system

Operation steps

(1) Turn on the computer and the instrument.

(2) Wash the 16 microspots of the Take3™ plate using Milli-Q and remove excess water by gently wiping with Kimwipes.

(3) Open the BioTek software and select "nuclear acid/protein measurement".

(4) Pipette 5 μL control buffer into the first microspot. This will be the "blank". Load 5 μL sample into individual microspots as "samples".

(5) Put the cover of the plate gently on top of the microspots and move the whole plate into the Multi-Volume Spectrophotometer System.

(6) Select the sample well on the software operation interface to mark the control and sample positions in the microplate.

(7) Press the OK button to measure the concentration.

(8) When the measurement has completed, the data will be recorded and displayed on the interface.

(9) Remove all liquid from the microspots, wash the plate with Milli-Q, and wipe excess water with Kimwipes.

(10) Turn off the instrument and computer.

Precautions

(1) When sampling with a pipette, the tip of the pipette should not touch the sampling groove.

(2) Control samples are needed.

(3) Even number of dots should be placed on the software operation interface.

(4) The instrument should be stored in a dry environment to prevent the growth of

mold and mildew inside the machine.

7. Gel imaging system

The gel imaging system is a device that digitizes the gel after electrophoresis and is usually equipped with UV and visible light lamps, a high-resolution CCD camera, and multi-functional analysis software. The machine uses UV to visualize nucleic acids or proteins in gels and the images are taken by the CCD camera. Special software is used to compare the positions and/or intensity of bands or protein spots for data processing, to qualitatively and quantitatively determine protein or nucleic acid expressions, colony plate counts, and other applications.

Operation steps

(1) Turn on the gel imager and allow the lamps to heat up for a short time.

(2) Turn on the computer and open the software.

(3) Place the gel to be imaged onto the stage and position it to the center.

(4) On the software operation interface, click PREVIEW to get a real-time image of the gel. You can adjust the gel position using this function.

(5) Select the filter suitable for the dye using the filter knob.

(6) Turn on the UV transmittance switch.

(7) Click LIGHTING in the software operation interface, and adjust the aperture, zoom and focus parameters until the image is clear.

(8) Click CAMERA and adjust the exposure time to get the best image.

(9) Click CAPTURE to take photos and save the files.

(10) After imaging, shut down the UV lamp and the camera system, clean the gel in the stage, and then process and analyze the images.

(11) Close the software and turn off the computer.

(12) Turn off the camera and gel imager.

Precautions

(1) Turn on the gel imaging system first, then open the software.

(2) Be careful not to contaminate the instrument, and do not touch the instrument door or computer with gloves that have touched the gel.

(3) The instrument door should be tightly closed when imaging the gel. Otherwise, the UV lamp may not work properly.

(4) Clean the inside of the machine once you have finished working with it.

(5) To extend the life of the lamp, please turn off the light source after visualiz-

ing or imaging the gel.

8. Electrochemiluminescence imaging system

Electrochemiluminescence (ECL) is widely used for protein quantification because it is very sensitive and has a low background signal. Recent development in the related fields, such as optical elements, luminophores, and electrochemical materials significantly increase the ECL signals. The ECL system consists of a bright field microscopy, a multiplex fluorescence detector, and a charge-coupled device (CCD) camera, which captures images of chemilumi-nescent signals (Fig. A-6).

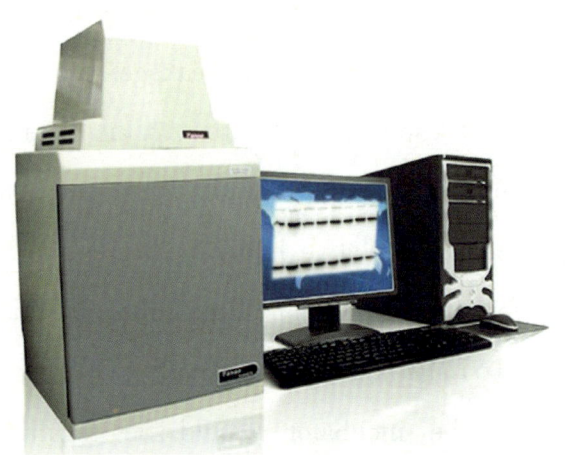

Fig. A-6　ECL image system

Operation steps

(1) Turn on the instrument and computer.

(2) Pre-cool CCD to −30 ℃.

(3) Mix the two substrates at a 1∶1 volume (generally 1 mL/membrane).

(4) Apply substrates to the membrane dropwise and cover it.

(5) After incubating for 1-2 min, pick up the membrane with tweezers, remove residual liquid, and lay it on the imaging tray.

(6) Preview the position of the membrane on the software operation interface, and adjust accordingly.

(7) Adjust the focus in the bright field mode, take pictures, and save files.

(8) Adjust the exposure time in the dark field mode, take pictures, and save files.

(9) Merge the bright field and dark field images into one and save.

(10) Once you have finished imaging, turn off the instrument and clean the stage.

Precautions

(1) Open the instrument and pre-cool the CCD to −30 ℃ before use.

(2) When the membrane is placed on the stage, it should be placed slowly and gently to avoid generating bubbles under the membrane.

(3) Clean the stage after use, and keep the machine dry.

9. CO_2 cell culture incubator

A cell culture (CO_2) incubator, creates a controlled environment with constant temperature, CO_2 levels, and humidity conditions conducive for cells/tissues to grow. This instrument has a CO_2 sensor that detects internal CO_2 level and transmits this information to the control circuit, solenoid valve, and other control devices. If the CO_2 concentration is low, the solenoid valve opens and CO_2 gas enters the chamber until the CO_2 concentration returns to preset levels. At this point, the solenoid valve is closed, the CO_2 supply is cut off, and internal CO_2 levels reach a stable state (Fig. A-7).

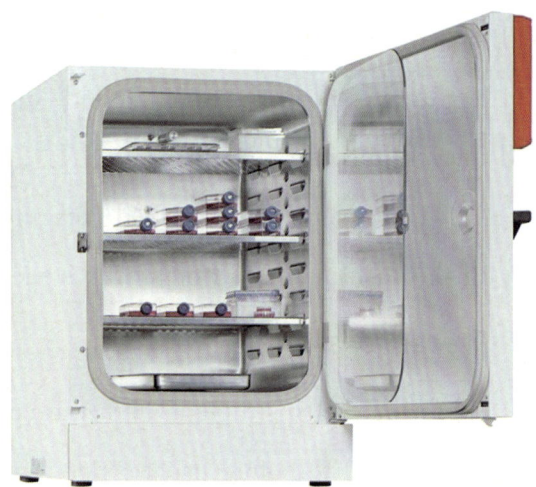

Fig. A-7 CO_2 incubator

Operation steps

(1) Turn on the incubator.

(2) Open the connection valve between the incubator and the gas cylinder.

(3) Press MODE to move the cursor to SET, press the arrow until the temperature information is displayed. Press the up or down arrow to set the required temperature value and press ENTER to save the set value.

(4) Press the left or right arrow until CO_2 concentration is displayed, press the up or down arrow to set the required CO_2 concentration, and press ENTER to save the set value.

(5) Press MODE to move the cursor to RUN position, and then start the machine.

Precautions

(1) Keep the air in the incubator clean and disinfect the incubator regularly.

(2) The gas supply should be pure and compatible to the equipment.

(3) Do not increase the inward gas pressure too much as this may burst the pipe and damage the detector.

(4) Close the incubator door when not accessing it to prevent gas leakage.

(5) Always pay attention to the amount of sterilized deionized water in the tank to maintain the relative humidity inside and to prevent the evaporation of the culture media.

(6) The temperature and CO_2 levels inside the incubator should be regularly calibrated.

(7) Each control switch should be turned off before shutting down the machine.

10. Flow cytometry

Flow cytometry (FCM) is a fast detection method that can measure many parameters and collect a large amount of data. This data can be comprehensively analyzed using a number of different analysis methods. FCM has been widely used in immunology, oncology, hematology, cell biology, biochemistry, cytogenetics, clinical medicine, botany, marine biology, environmental science, pharmacy, and other biological fields.

In FCM, properly dispersed cells are suspended in an isotonic solution, which allows the cells to pass through the detection point as single cells and produce signals orthogonal to the laser. When the laser hits fluorescently labeled cells, they emit scattered light and fluorescent signals, the latter of which is emitted in a 360° pattern. The scattered light does not come from the fluorescence label, but is inherent to the cell, and reflects the size and internal structure of the cell. Therefore, the intensity of the scattered light can be used to preliminarily group cells.

The scattered signal is divided into forward-scattered light (FSC) and side-scattered light (SSC). FSC refers to the scattered light emitted along the path of the laser, and this value is proportional to the diameter of the cell. In other words, the larger the cell is, the stronger the FSC intensity. SSC refers to light emitted perpendicular to the path of the laser and is sensitive to the refractive index of the plasma membrane, cytoplasm, and nuclear membrane. SSC intensity linearly correlates with cell granularity and can provide information about the fine structure and particle properties inside the cell. Therefore, the more complex the cell structure is, the greater the SSC intensity. Both FSC and SSC have the same excitation and emission wavelengths. A combination of these two parameters can be used to distinguish cells of different sizes and intracellular constituents.

Fluorescent substances emit light at a longer wavelength than the wavelength it was excited with. Substances that can produce fluorescence upon excitation are called fluorescent substances or luciferins. Cells excited by the laser generates scattered light and fluorescence, which are then separated by optical elements such as lens, binary mirror, and filter, and finally transmitted to the corresponding detectors. Each detector can only detect a certain wavelength range, which means that each detector detects only a limited type of scattered light or fluorescence.

In summary, in FCM, single cells are delivered to detection points under pressure in sheath fluid. When cells pass through the detection points in a single file, the laser hits the cells and emits scattered light. At the same time, the fluorescein on the cells is stimulated by the laser, which emits fluorescence. Both the scattered light and fluorescence are received by the corresponding detectors and converted into electrical signals that are transformed into digital signals for computer analysis. If the target cells are detected, they can be sorted by electrostatic attractions or repulsions (Fig. A-8).

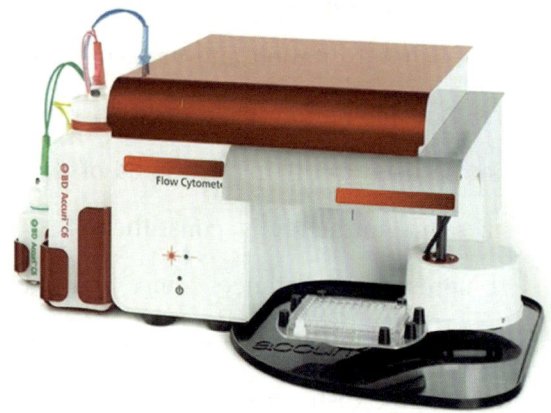

Fig. A-8 Flow cytometory

Operation steps

(1) Add the sheath liquid and empty the waste liquid bucket. Turn on the instrument according to the boot sequence, and allow the machine to warm up for 5-10 min.

(2) Remove air bubbles from the tubing and rinse the needle with ultrapure water at high speed for about 2 min.

(3) Open the software for data acquisition and analysis.

(4) Start with the control sample and set the voltage for each parameter.

(5) Set the voltage for FSC and SSC so that the cell population appears at the center of the FSC-SSC scatter plot.

(6) Draw a polygonal gate (R1) to select the cell population. Only the cells in R1 will be analyzed for GFP transfection efficiency.

(7) Adjust the voltage of GFP. This population or peak should appear on the left side of the GFP-SSC scatter diagram and the single-parameter histogram of GFP. Then, on these graphs, draw the R2 and R3 gates to select the SSC and GFP populations.

(8) After the voltage and gates of each parameter have been set, run the samples, and acquire data from at least 10,000 target cells per sample.

(9) Clean the instrument according to the manufacturer's instructions. After cleaning, exit the software and turn off the instrument.

Precautions

(1) The sample must be filtered with 40 μm-filter membrane before loading to prevent pipeline obstruction.

(2) After the experiment, the pipes must be cleaned with sodium hypochlorite and Milli-Q sequentially to prevent pipeline obstruction by pollutants and crystals.

(3) The equipment can be shut down only after the completion of automatic cleaning.

(4) It is necessary to empty the sheath liquid bucket to prevent bacteria growth when the equipment is not used for a long time.

(Written by Chen Xiaoxia and Zhang Yan)

Section 2 Preparation of Commonly—Used Solutions

① 50 × TAE Buffer (1 L)

Tris ·················· 242 g Tris

Acetate ·················· 57.1 mL

500 mM EDTA (pH 8.0) ······ 100 mL

Add deionized water to make a final volume of 1 L. Use deionized water diluted 50 times before use.

② 4 × DNA Lysis Buffer (250 mL)

200 mM Tris-HCl (pH 8.0)

·················· 50 mL

1 M EDTA (pH 8.0) ········ 100 mL

10% (W/V) SDS ············ 50 mL

Add deionized water of 50 mL to make a final volume of 250 mL.

③ LB Liquid Medium (pH 7.2, 1 L)

Tryptone ·················· 10 g

Yeast extract ·················· 5 g

NaCl ·················· 5 g

Add deionized water to make a final volume of 1 L,

Adjust pH with NaOH to 7.2.

Steam sterilization at 121 ℃ for 30 min.

④ LB Solid Medium (pH7.2, 1 L)

Tryptone ·················· 10 g

Yeast extract ·················· 5 g

NaCl ·················· 5 g

Agar ·················· 15 g

Add deionized water to make a final volume of 1 L.

Adjust pH with NaOH to 7.2.

Steam sterilization at 121 ℃ for 20 min.

⑤ SOC Medium (pH 7.0, 1 L)

Trypton ·················· 20 g

Yeast Extract ·················· 5 g

NaCl ·················· 0.5 g

250 mM KCl ·················· 10 mL

Add deionized water to make a final volume of 1 L.

Adjust pH with NaOH to 7.0.

Steam sterilization at 121 ℃ for 20 min. Cool to 60 ℃ or below, add 2 mol/L sterilized $MgCl_2$ (5 mL) and 1 mol/L glucose solution filtered with 0.22 μm filter (20 mL).

⑥ 100 mM IPTG (10 mL)

IPTG ·················· 0.24 g

Add sterilized water to make a final volume of 10 mL.

Sterilized with 0.22 μm-filter.

⑦ 40 mg/mL X-Gal (10 mL)

X-Gal ·················· 0.4 g

Add DMF to make a final volume of 10 mL.

⑧ 100 mg/mL Ampicillin (10 mL)

Ampicillin sodium ············· 1 g

Add sterilized water to make a final volume of 10 mL.

⑨ RIPA buffer (500 mL)

1 M Tris-HCl (pH 7.4) ········ 25 mL

5 M NaCl ················ 15 mL

NP-40 ·················· 5 mL

10% SDS ················· 5 mL

Add deionized water to make a final volume of 500 mL.

⑩ RIPA Buffer with PMSF (5 mL)

RIPA buffer ··············· 5 mL

100 mM PMSF ············· 50 μL

Prepare currently and keep on ice, avoid light.

⑪ 30% (W/V) Acrylamide (250 mL)

Acrylamide ··············· 72.5 g

BIS ··················· 2.5 g

Add deionized water to make a final volume of 250 mL.

⑫ 5 × Tris-Glycine Buffer (1 L)

Tris ··················· 15.1 g

Glycine ················· 94 g

SDS ··················· 5 g

Add deionized water to make a final volume of 1 L.

Use deionized water diluted 5 times before use.

⑬ 1.5 M Tris-HCl (pH 8.8, 250 mL)

Tris ··················· 45.4 g

Add deionized water to make a final volume of 250 mL.

Adjust pH with HCl to 8.8.

⑭ 1.0 M Tris-HCl (pH 6.8, 250 mL)

Tris ··················· 30.3 g

Add deionized water to make a final volume of 250 mL.

Adjust pH with HCl to 6.8.

⑮ 1 × Transfer Buffer (1 L)

Glycine ················· 2.9 g

Tris ··················· 5.8 g

SDS ··················· 0.37 g

Methanol ················ 200 mL

Add deionized water to make a final volume of 1 L.

⑯ 10 × TBS (pH 7.6, 1 L)

    Tris ·············· 24 g

    NaCl ·············· 88 g

Add deionized water to make a final volume of 1 L.

Adjust pH with HCl to 7.6.

⑰ 1 × TBST (1 L)

    10 × TBS (pH 7.6) ······ 100 mL

    Tween-20 ·············· 1 mL

Add deionized water to make a final volume of 1 L.

⑱ 10% (W/V) SDS (pH 7.2, 100 mL)

    SDS ·············· 10 g

Add deionized water to make a final volume of 100 mL.

Adjust pH with HCl to 7.2.

⑲ 10% (W/V) AP (10 mL)

    Ammonium persulfate ······ 1 g

Add deionized water to make a final volume of 10 mL.

⑳ Block Solution [5% (W/V) non-fat dry milk, 100 mL]

    1 × TBST ·············· 100 mL

    non-fat dry milk ·············· 5 g

Dissolve the non-fat dry milk thoroughly.

(Written by Chen Xiaoxia)